S r

The beginner's guide on how to design and install a photovoltaic system for your home, cars, vans and boats

By Travis Newman

Table of contents

Chapter I

Introduction

In recent years, the growing energy needs and increasing environmental changes are under consideration. Alternatives of non-renewable and polluting fossil fuels have to be investigated.

Importance of Solar energy

No one can deny the importance of solar energy. Among all the other options for non-renewable energy, the most suitable one is solar energy. Solar energy is produced directly by the sun as the radiations of the sun are converted into electrical energy and collected elsewhere, usually the Earth. This process generates heat and electromagnetic radiation through a thermonuclear process. It converts about 650 000, 0001 tons of hydrogen to helium every second, which is quite a significant amount. The heat always remains in the sun and is influential in maintaining the thermonuclear reaction. The visible light, infra-red light, and ultra-violet radiation, which are the electromagnetic radiation which stream out into space in all directions. Of the total radiation produced only a minimal amount reaches the Earth. Solar energy and power can be used in multiple ways. We'll discuss it in later chapters.

The indirect source of Energy

Another critical point is that the radiation that reaches the Earth is the indirect source of almost all types of energy that we use today. The exceptions, in this case, are geothermal energy, nuclear fission, and fusion. The fossil fuels that are also considered as renewable energy source owe their origins to the sun. In fact, they were once living plants and animals whose life was dependent upon the sun.

Nowadays, much of the world's required energy can be supplied directly by solar power. Moreover, it can also be provided indirectly. The practicality of doing it will be examined, as well as the benefits and drawbacks.

Two main components of a solar energy generator

There are two primary components of a solar energy generator. These are the two main components which are required to have a functional solar energy generator in the production of electricity through solar energy. The two parts are known as a collector and a storage unit. Let's discuss these parts briefly.

Collectors

The collector has to collect the radiation that falls on it and convert a fraction of it either into electricity and heat energy or heat alone. At certain times, only a minimal amount of radiation will be received in the storage unit. During heavy cloud cover or at night, for example, the amount of energy produced by the collector will be quite small.

Types of collectors

In general, solar energy systems have three types of collectors and many forms of storage units. The three types of collectors are focusing collectors, flat-plate collectors, and passive collectors.

1. Focusing Collectors

The focusing collector is one of the main types of collectors. Focusing collectors are specific for maximizing the radiation falling on the focus of the collector. They are necessarily flat-plane collectors with optical devices. These are not so popular and are currently used only in a few scattered areas. An example of this type of collector is a solar furnace.

The negative aspect of focusing collectors

A negative point of focus collectors is that they lose some of the radiation that the flat-plane panels do not. Radiation reflected off the ground will be ignored by focusing collectors, but flat-plane panels uses it. Another problem with focusing collectors, in general, is due to temperature. The focusing collectors, by their very nature, need safeguards to protect their silicon components, and also they can create much higher temperatures.

2. Flat-Plate collectors

Another significant type of collector is flat-plate collectors. Flat-plate collectors are arrays of solar panels that are arranged in a simple plane, and they are the collectors that are the more commonly used type of today.

The solar energy systems can be near to any size, and the output that is provided is directly related to a few variables, which include size, cleanliness, and facing. The collector panels in solar energy systems have automated machinery that keeps their angle towards the sun.

3. Passive collectors

The third collector, which is passive collectors are entirely different from the other two types of collectors. The function of passive collectors is to absorb radiation and convert it into heat naturally (without being explicitly designed for the purpose).

Although all the objects have this property to some extent, rarely are there some objects (like walls) that are able to produce enough heat to make it worthwhile. The passive conductor has a natural ability to convert radiation into heat, enhanced in some way or another, as it can, by being painted black; and transferring the heat in a system for a different location is generally added.

Storage Unit

The storage unit can hold a tremendous amount of energy produced during the periods of maximum productivity, and the power is released when the productivity drops. In practice, methods of collecting and storing solar energy vary on different aspects, depending on the uses planned for the solar generator. It depends on your need and requirement to choose any of these multiple options and aspects.

Tasks that consume power

As for most of the purposes, people use energy, but there are only a few general tasks that consume most of the power. These tasks include heating, transportation, cooling, and the generation of electricity. Solar energy supports all four of these tasks with different levels of success.

Heating Purposes in business

For heating purposes in the industry, solar power is best suited. Almost no energy transformation is required in solar heating, so it has very high efficiency. Heat energy can be stored in a packed bed or a liquid, such as water.

Packed bed

A packed bed is a container that is usually filled with small objects that can hold heat with air spaces between them. Heat energy can also be stored in heat-of-fusion units or a phase changer.

Advantage of heating through the solar system

Generally, a building that's heated by solar power has an advantage. The advantage is that it will have its water heated by solar energy as well. As passive collectors are an integral part of the building, for buildings to take advantage of passive collectors, buildings must be created with solar heating in mind.

Forms of Passive collectors

These passive collectors can be used in different ways. Among all the kinds of passive collectors, the most basic type is the incidental heat trap. It allows the maximum amount of possible light inside through a window.

Solar Pond

Another main and the most significant form of the passive collector is the solar pond. The main purpose of this solar pond is to serve as an energy regulator for a building, which is quite an important task. This pond absorbs solar energy and converts it into heat during the day. This heat can quickly be taken into the building, or if there is more than enough heat already, heat can be dumped into the pond as well.

Cooling through solar energy

Solar energy and power can be used for other things; among them, the most common uses of solar energy today is cooling. Solar cooling is more expensive than solar heating, so it is almost never seen in private homes.

Process of cooling

Solar energy is also used to cool things by phase-changing through the process of evaporation. A liquid can be converted into gas through heat, and then the gas is forced into a lower pressure chamber. The temperature of a gas is compared to the pressure containing the chamber, and even all the other things being held equal, the same gas under a lower pressure, will have a lower temperature.

Conversion of solar energy into electricity

One of the main focuses of solar energy is the direct conversion into electricity. Most of our tools use electricity that is designed to be driven by it. So, if you are capable of creating electricity through solar power, you can run almost anything with solar energy. Flat-plate collectors or focusing collectors are the solar collectors that convert radiation into electricity. To perform such a task, collectors contain silicon components, which are photovoltaic cells. This type of cell does not have any authentic explanation as it has yet to be field-tested. If the cell maintains its efficiency in the uncontrolled environment of the outside world, and if it does not tend to break down, it will be economical for power companies, which can then built solar power facilities. So, we can say that the efficiency of the cell can affect the working of the solar panel to a great extent.

Least Suited facility; Transportation through solar power

The least suited facility to solar power is transportation. Most of the time, large, relatively slow vehicles like ships could power themselves with large onboard solar panels, but these do not support small, always turning vehicles like a car. The only possible way in which a car could be completely solar-powered would be through the use of a battery that can be easily charged by solar power at some stationary point and then later loaded into the car. The invention of electric vehicles can be partially powered by solar energy, which is highly available now. Still, it is unlikely that solar power will provide the world's transportation alternative shortly.

Benefits of solar power as compared to fossil fuels

Solar power has more advantages over fossil fuel, when comparing renewable energy sources. The first fact is that it is never going to run out. The second is its positive effect on the environment. The burning of fossil fuels harms our environment, as it introduces many harmful pollutants into the atmosphere and contributes to environmental problems like global warming and acid rain; solar energy doesn't create any pollution. The destruction of many acres of land to establish a fossil fuel energy plant is mainly due to its required fuel. However, no property will be destroyed if a solar energy system is incorporated into every home and business.

Moreover, the use of silicon is also beneficial as the primary element of the construction of solar panels. Silicon is the second most common element on the planet, so there won't be any environmental disturbance caused by the creation of solar panels. Solar power certainly can be produced on a gigantic scale, too. The only thing that affects the working of solar energy systems is the change of temperature and humidity near a solar panel.

The perfect resource for the future

Solar power is the most promising source of all the energy sources available nowadays. If we consider it numerically, solar energy is capable of producing the raw power required to satisfy the entire planet's energy needs. Thus, the use of solar power can help us in multiple ways at an affordable cost. Another factor that is the effect on the environment; so environmentally, it is one of the least destructive of all the sources of energy. Thus, practically, it can be adjusted everywhere to power nearly everything, except transportation because transport requires a little adjustment. We can conclude that solar energy is a perfect resource for the future.

Chapter 2

What is Solar Power?

Source of Energy

According to scientific research, the Earth captures a large amount of solar energy, 173 trillion terawatts to be exact, which is ten thousand more power than the whole world population utilizes. It confirms the fact that the Sun is the most abundant source of energy on Earth, and no one can deny it.

A most suitable source of energy

Few countries have complete entrance to fossil fuel-based energy resources, which can lead to global governmental and budgetary instability. However, the most suitable choice is solar energy, which is a renewable resource. It grants an endless, regular supply through time. Solar energy is a green source of energy. It does not discharge pollutants during the process of energy production.

Power is extracted from the usage of resources, either physical or chemical, especially to work machines by providing light and heat. Solar means anything related to Sun. So, solar energy is the energy taken from the Sun.

Definition of Solar Power

"**Solar power** is the luminous light and heat taken from the Sun that has been domesticated by man since classical times, practicing a range of ever-evolving technologies."

Undoubtedly, solar power is one of the best and useful powers as the light and heat coming from the Sun are limitless. Solar energy is also a renewable source of energy, i.e., and it can be used again and again. Energy taken from the radiations of Sun is solar energy. It is used in various ways on Earth for serving different purposes. Let us discuss some of the technologies of solar energy.

Solar Energy Technologies

Solar Energy Technologies are the uses of solar energy for practical applications and purposes. Almost all other renewable energies derive their powers from solar radiation.

Active and Passive uses of solar energy

Solar energy is used as a passive as well as an active source of energy. When solar energy is used directly, it is called active use, and when it is used indirectly, it is called passive use of energy. For example, sunlight comes through a window and heats a room is a passive use of energy, while the conversion of sunlight into electricity is active use of energy.

The solar energy technologies are also classified as active and passive solar energy technology. An active solar technique is the use of photovoltaic modules (also called photovoltaic panels) and solar thermal collectors, which convert sunlight into beneficial outputs.

The passive solar technique is directing a building towards the Sun, selecting materials with favorable thermal mass or light dispersing properties, and designing spaces that naturally circulate air.

Solar thermal technologies

Solar radiations are converted into electricity and heat. The techniques converting sunlight into heat energy are called solar thermal technologies.

How to convert solar radiation into light energy?

Photovoltaic cells (PVCs) are used to convert solar radiation to electricity. It is done by stimulating electrons in silicon cells, which then capture the photons of light from the sun-rays. This electricity is then used to provide renewable energy for your home or business.

The photons irradiated from the Sun and reach the Earth must be accumulated, transformed into a convenient arrangement, and then conveyed to an electric device or the electrical grid; both can be used to generate solar energy. Displays of photovoltaic cells are commonly used to collect energy from the Sun and convert it into electricity.

Renewable sources of energy have passed coal, which was the world's largest source of electricity generation. About 30% of that capacity is because of silicon solar cells.

Working of silicon cells

A silicon cell consists of thin strips of metallic electrodes on either side. The power is extracted by electrodes from the core of the solar cell, which sends it to the external circuit.

The silicon cell is like a sandwich. Just like the filling of the sandwich, the internal working of the cell is most captivating. It is where the main action is done - the photons captured from the Sun are converted into electricity.

Layers of silicon

The solar cell consists of two layers of silicon - negative and positive silicon, which are termed as n-type and p-type silicon.

Dopants

Creating positive and negative poles in silicon is done by impregnation of elements known as dopants. Dopants are elements that can replace some of the silicon atoms in the crystal structure, manipulating the number of electrons available in each layer.

For example, phosphorus is employed to create n-type silicon and boron is used to develop p-type silicon. Phosphorus has one electron more than silicon. When phosphorus is added to silicon, the extra electron of phosphorous is very weakly attached to it. The weakly attached electron can move freely, hence charging the silicon negatively.

At the juncture of the p- and n-type silicon, the positive electron holes, and the electrons combine. This movement gets you a slightly positive charge on the n-type silicon and a slightly negative charge on the p-type silicon at the juncture of both.

Photons from the Sun move amidst the layers of the prime electrode and ignite silicon atoms in the crystal structure. The coinciding photons give some of the silicon electrons enough energy to escape from their parent silicon atom.

The "free" electrons move to and gain within the n-type silicon. Once free electrons have gained in the n-type silicon, the free electrons are stimulated to work. The electrodes must be connected through an external circuit to use the energy of free electrons. Electrons move through the electrodes and the external electric circuit from the n-type to the p-type. The p-type silicon acts as an electron ewer. It keeps the electrons flowing; otherwise, they would remain static.

This flow of electrons creates the electrical current to power appliances or charge batteries while the Sun isn't beaming. This technology is used to utilize the light of Sun and to convert it into useful energy sources, there are a variety of ways in which this conversion can take place. Based on that, we have a few types of solar energy systems:

1. Photovoltaic systems

These systems include the use of photovoltaic cells or solar panels found on the rooftops or spacecraft. These significant photovoltaic cells are made up of semiconductor materials. The conversion starts when sunlight hits these cells, and the electrons in the cells get loose from their atoms. As these electrons move freely through the cell, they give rise to electricity.

2. Solar water heating system

As the name "solar water heating system" suggests, the method involves the use of the Sun to heat water and generate electricity. This system has two parts - a storage collector and a storage tank.

The collector is set on the roof facing the Sun. In it are small tubes in which fluid runs through. When the collector gets heated, it heats the fluid moving through the tubes. The tank is then used to store the hot liquid.

3. Concentrated solar power

At large scale electricity production, this method is used, which involves the positioning of mirrors over vast areas. Firstly, the concentrated solar plants convert Sun's energy, which falls onto the mirrors, into heat. Then it converts that high-temperature heat into electricity, which can be utilized later on in industries, companies, etc.

Use of Solar energy

Solar energy nowadays can be used in multiple ways. It depends on human beings to decide how they wish to utilize it. One can use it in different ways and numerous walks of like. A large number of people are already getting benefits from solar energy. The use of solar power has dramatically intensified in recent years, and nowadays, it is used in new and innovative ways. Some of these use include:

1. Solar-powered transportation

Solar power can be used for transportation purposes. Nowadays, buses, subways, railroads, planes, and cars are all being powered through solar energy by the use of photovoltaic cells. Solar transit is now in the trend as a renewable energy source.

2. Solar tech

The tech-savvy world that exists today has put forth the use of charging our mobiles, tablets, or any other wearable appliances through sunlight. Now there are solar-powered flashlights that are charged by exposing it to sunlight. This day by day development in the world is urging people to use solar power for charging their everyday electronics.

3. Solar lighting

Solar lighting is an efficient way to store light and use it at night. There is no complexity to these lights as they are wireless, and also they are cheap and improve your outdoor decor. Although they are not that much in the market as far as solar panels are, they are still making their mark alongside LED flashlights and bulbs.

4. Solar heating

Solar heating is also an effective way to heat our homes without using solar panels. Solar space heaters convert Sun's light into thermal energy by using air or water as a medium, whereas solar water heater uses water to generate electricity. These systems can either be passive (using natural circulation) or an active system that uses pumps to generate water. Moreover, installing a full solar panel system is costly, but that is not the case with thermal power as you can expect a 5 to 10 percent return.

5. Rooftop Solar

The cliché and old use of solar panels that are still hanging around today and used all over the world by people. This solar paneling system helps you to cut down on your massive electricity bills. The only way to improve your carbon footprint and electric bill is with solar panel installation on your roof.

6. Charging batteries with solar power

Any device that uses a battery can be charged with solar energy by a photoelectric solar panel system whose purpose is to generate DC that is used in batteries.

7. Road signs

A solar panel can even be used to light up road signs and message boards. It is charged in the say, so that the sign is lighted up throughout the night.

In this pollution filled world, the use of solar energy is a great initiative that conserves energy, is renewable, prevents any harmful gas emissions, and is nowhere costly. Whether you use it on an industrial level or in households, it is useful. The Sun has a lot to give us.

Chapter 3

Solar energy history

Solar panel techniques are relatively new technology (as compared to that of fossil fuels). However, people are using the sun's energy for multiple tasks from a long time ago. The ancient people have already used solar power in the 7th century B.C. The concept of using solar energy was a common practice for older people and Greeks thousands of years ago. This concept of solar power still has considerable potential for growth in the future. Some phases of the journey of solar energy are described below.

The Ancient time (lighting up the fire)

In the early times, there were no methods and proper ways to light up the fire, so the Ancient people from the 7th century used the energy from the sun through a magnifying glass to set wood on fire for cooking. Later, in the 3rd century B.C, the Greeks and Romans used solar power. They used it by using the reflection of the mirror.

Another early use of solar energy that is still popular today in many modern houses was the concept of sunrooms in the buildings. With the help of these sunrooms, the sunlight was directed into one concentrated area through massive windows to capture its natural heat. Later in 1200 A.D, the Native Americans used this concept of sunrooms to capture the natural heat of the sun for winter months. The people at that time tried different methods to catch the energy from the sun for different purposes, like a heating water tank or to burn enemy ships. So, we can say that the energy from sunlight has been used in various ways throughout history.

The time of 1700 A.D and 1800 A.D

Later in the 1700s and 1800s, the scientists and researchers of that time got success in their project of using the sunlight to power ovens for long trips and tours. They also utilized the power of the sun to produce solar power stream boats.

In 1839, scientist Edmond Becquerel who was credited for the invention of solar cells, discovered the photovoltaic effect by the experiment of electrolytic cells with sunlight. During this process, he realized that more electricity was produced if the cells are exposed to sunlight. In 1876, a group of various scientists researched the photovoltaic effect and discovered that selenium creates electricity when exposed to sunlight.

After few years, in 1883, American inventor Charles Frits produced the first working solar cells made from selenium wafers. This selenium cell was a significant component of the technology being used today. In a way, many scientists played a part in solar cell invention.

The time of the 1900s century

In 1908, scientists produced solar collector devices to absorb solar energy, and these devices were similar to those in use today. At that time, the people got awareness about solar energy from the publication of the photovoltaic effect on paper and the publications of books about solar houses.

In 1956, the first solar cell was available for commercial use. These cells were costly and were far from the reach of everyone. At that time, the primary solar cells were used in mini toys and radios to check productivity.

In the 1960s, the solar cells were used to power the satellites in space by USA Scientists. At that time, the people were looking for the cost deduction of the solar cells so that it could be available for everyone to use this beneficial technology. So in the 1970s, they got succeeded in discovering the low-cost solar cells. With this discovery, the price was brought down from 100 dollars per watt to approximately 20 dollars per watt. It was the beginning of the commercial use of solar panels, and from this time scenario, the solar panels were not only specified to the large companies, but they were available for domestic and public use.

From this period, the use of solar energy through solar cells was established and modified. With low cost, they were used in different appliances, to power houses, for telecommunication, and many other projects.

The 2000's

From the beginning of this twenty-first century, scientists were working and focusing on making solar panels more efficient to operate and reliable. The solar panels were already available at low costs, so the primary purpose of working on it for the scientists was to make it more efficient.

As technology changes with time and becomes more accessible to everyone, the goal was to make solar power less expensive, like other power resources for everyday use. In this time, the solar panels were modified to be used in transportation and for communication systems.

A summary of the history of solar power

The history tells us that solar energy was used by humans thousands of years ago (in the 7th century B.C). The people at that time used sunlight to light fires through a magnifying glass. Later in the 3rd century B.C, people used sunlight to light up the torches for religious ceremonies. As time passed, people used this concept of reflection of the sun to create sunrooms in buildings to capture the sunlight to a concentration area.

In the 17th and 18th centuries, scientists and researchers succeeded in using solar energy to light up the ovens for long tours. In this time, the photovoltaic effect was discovered due to which scientists produced the first-ever solar cell to use solar energy for the production of electricity. So, solar energy is not a new thing; it was used by the ancient people thousands of years ago as a regular practice for them for various purposes and it's usage developed throughout the years to what we know it as today.

The invention of solar panels

Solar panel technology was developed through the hard work and contribution of various scientists and researchers. Firstly, the scientist, Edmond Becquerel, discovered the photovoltaic effect by the experiment of the electrodes with sunlight, and concluded that when these electrodes exposed in front of the sun, they produce electricity. Other scientists and researchers used the concept of this effect, and in 1883, scientist Charles Frits discovered the solar cells.

These selenium cells were not much efficient for the productivity of electricity, so later on, the solar cells made up of silicon were discovered by a scientist Calvin fuller in 1953. These cells produced much power to run small electrical devices like toys and radios. From this time, the beginning of a new era for solar cells began.

In the 1960s, the solar cells were used in satellites and spaceships, and with time, technology changed, and scientists tried to focus on the efficiency and to lower the cost of solar cells for commercial use. From the

1970s onwards, the solar panels became able to be used in communication systems and transportation and for other large and small scale projects. The cost also lowered over time so that this technology could be available for everyone to use.

Awareness of Solar technology

In the 19th century, the scientist, Albert Einstein played a vital role in bringing the attention of the world towards solar energy and its importance. In 1905, he published a paper about the photovoltaic effect of light and how light carries the energy and how we can utilize this energy. This research generated more attention towards solar power.

As time passed, scientists worked together to generate silicon solar cell, which was more practical; this was the beginning of the new era for solar power. The solar cells were being used to power toys, radios and other small devices and were commercially used. This made people more aware of solar energy and its practical results.

From this time to the 1970s, solar energy became a general and low-cost energy to be used to light up the houses and other electrical projects. Solar panels were used as an alternative of the usable energy in the 1970s when the U.S. experienced a shortage of oil and the country was struggling economically. The lack of requirements and needs motivated scientists to invent solar panels. During this time, the solar panel was installed onto the White House roof. It was a statement to make energy through solar power more reliable and to spread awareness to the people about solar energy.

As the time passed, the scientists made improvements in lowering the cost and efficiency of solar panels so that more people took an interest in solar power as compare to other ways. Solar power then became a realistic way for everybody to generate electricity for their houses and business and other appliances in daily life. The continuous modification and contribution of various scientists of the 19th-century helped in making solar power spread everywhere, with lower cost and high efficiency for utilization in appliances for everyday use.

From the 1980s, the cost of solar cells was lowered by 10% every year with efficiency. This further motivated the widespread use of solar power.

Efficiency and price over the years

The first solar cell, which was discovered in 1839, was one percent efficient, and its cost was 300 dollars per watt, which was too expensive and less productive. In 1954, when silicon cells were operated in the lab, scientists were able to increase this to 11% efficiency of cells. The efficiency then increased with time, for in 1960, 14% efficiency was achieved. This efficiency upgrade helped scientists to use solar cells for space programs. It increased productivity and lowered the cost up to 100 dollars per watt. It was a great initiative taken by scientists at that time.

As time went by, the improvement and modification of solar cells made it more efficient and lower cost and became a source of energy for everyone to light up their houses and for their business.

Conclusively, ancient humans introduced the solar power concept. As the technology improved throughout the centuries, the solar power systems also upgraded and became more accessible in all aspects of life. This was made possible through the hard work and contribution of various scientists.

Chapter 4

The Advantages of Solar Energy

There are numerous advantages of solar energy. Let's have a brief and clear look at some of the benefits of a solar energy system.

Clean Energy Source

Solar energy offers a renewable, domestic, and clean energy source. It reduces dependency on oil, coal, and natural gas for electricity production. It is also a crucial component of sustainable energy. Solar energy systems provide notable environmental benefits.

Protecting the environment

The Sun's productive power offers an infinite source of energy that does not destroy the landscape or harm the ozone layer. The residential solar energy systems represent an investment in the future of the planet by protecting the environment and conserving non-renewable energy sources.

Diversity in the application of solar energy

There are diverse applications of solar energy. The photovoltaic and thermal energy can be generated through solar energy. Solar energy can be used to distill water in the regions where there are limited clean water supplies, to produce electricity in areas that do not have access to the energy grid, and to provide power to satellites in space.

Construction of buildings

Solar energy can also be applied to the materials used for the construction of buildings. The introduction of solar architecture like compact proportion (ratio of surface area and volume), thermal mass, orientation, and selective shading are all tailored to a specific climate and particular environment, so the building remains in a very comfortable temperature range. Ventilation systems, solar lighting, and heating are all combined together in the solar design as well as transparent solar energy windows.

Slightest negative impact

The production of solar energy is considered safer than the production of power from fossil fuels, for humans to operate and to produce dangerous amounts of radiation. Solar energy systems have the slightest negative impact on the environment compared to any other energy source.

Purifying drinking water

Solar energy can even be employed to treat or clean drinking water, which is a significant public health benefit in the developing world. Solar energy is a lot cleaner when compared with other energy sources. These systems have many significant advantages, like being cheaper and not producing any pollutants during operation, and being almost an infinite energy source when compared with fossil fuels.

No noise pollution

As the production of solar energy does not generate any noise pollution, so it becomes an important matter to take into account for installations of solar energy in urban areas. It also doesn't create any waste as it does not need any maintenance, and its lifetime is far longer than other energy-production systems.

Prevent Air pollution

Solar energy helps to get rid of air pollution. It is one of the ways through which we can prevent the effects of climate change by releasing fewer pollutants in the air and reducing carbon dioxide emissions. Solar energy reduces demand for fossil fuels, limits greenhouse gas emissions, and diminishes your carbon footprint. Cleaner air has a positive effect on our health.

Avoid health problems

Solar energy adoption significantly reduces the number of nitrous oxides and sulfur dioxide emissions, all of which can cause health problems.

Helps to reduce water consumption

As we know that all manufacturing processes require some water, also water is used to make solar panels, but the total amount of water needed to generate solar electricity is fiercely less than more conventional electricity sources. Therefore, it helps reduce water consumption. So, with solar energy, there's no risk to local water resources, and their operation does not burden local supplies by competing with agriculture, drinking systems, and other vital water needs. It surely helps solve the water scarcity issue.

Protected against disruptive effects

Solar energy facilities can be so broadly spread out. As they are made up of many separate devices, therefore, they are better protected against disruptive effects like storms, which can eliminate power to large populations by damaging just one generator or transformer station. Solar energy technologies like solar panels and home batteries are becoming more accessible, resilient, inexpensive, and affordable. They also provide energy and peace of mind, security for residents when the power dies out, or the next storm strikes. Also, it is well known that many solar technologies use less water than other fossil-fuel or nuclear power plants. In fact, they may also be more resilient in the face of dry spells. Solar energy system replenishes with energy generated from the Sun and eliminates the inconvenience, uncertainty, diesel-powered generator, and expense of refueling of gas.

Interchangeable

Solar energy is highly interchangeable; it is composed of many individual installations that can be linked together. The solar energy can be implemented at many scales - from distributed generation through rooftop solar panels to a utility-scale thermal plant. Solar technology also tends to be low-maintenance. Since solar energy doesn't require any external supply to work, its maintenance and energy production costs are almost zero. Solar systems have tremendously localized solar setups; for the present, they can work well in rural or in developing areas where grid energy is unreliable, highly expensive, or unavailable.

Affordable source of energy

The solar energy technologies usually require a significant initial investment, but operational costs are low, and the fuel such as light and heat from the Sun are free. However, the latest technological advances point towards an unusual drop in the prices of the components required to manufacture them, which will translate into more efficient and affordable solar cells. Although there is an increase in technological improvements, expanding markets, and government subsidies and incentives, solar-technology costs have declined in recent years.

Job creation

Solar energy, in general, is considered more labor-intensive than the other energy production sectors, so it is more capable of supporting more jobs per unit of energy produced. The widespread solar energy systems promote local job creation and have a positive impact on the economy in the area. The increased production of solar energy can result in more than three times the number of new jobs.

Sunlight is a universally available resource of energy as compared to other potential resources of energy, though, of course, it varies geographically and seasonally in amount and intensity. Gaining on such a potentially productive domestic energy supply can reduce a country's dependence on foreign energy sources. It is also a distributed energy system which is better safeguarded from natural disasters.

Reduces Electricity Bill

Solar energy technologies savings can even accumulate on cloudy days since the Sun emits energy through clear and cloudy skies. Solar energy can also be employed in remote and rural regions. Solar power offers year-round productivity and savings, depending on their size, efficiency, and orientation relative to the Sun, even in cloudy and colder climates. Some solar energy systems generate more electricity than the home consumes. That can most certainly reduce your monthly electric bill to zero.

Improves Grid Security

The solar power system improves grid security as the grid is less defenseless to blackouts if many power plants are outstretched. A grid with a great discernment of solar energy has thousands of energy production centers that are widely expanded. It improves the security of the grid in case of overload, natural disasters, or human-caused disasters. The solar energy system can be connected to the electricity grid. Solar energy systems improve the security on the power grid against overloads or fires in transformer substations. Once an individual has the approval of the solar power system connected to the electricity grid, that person may have the option of exporting and selling their unused solar power back to their electricity retailer.

The local electricity loses some of the energy, around 3-5% during transportation and distribution, but if you have solar energy systems on the roof or in the yard, it notably reduces this distance, therefore enhance the efficiency of the solar power systems.

Simple and easy to install

Solar energy systems show ease and simplicity in the installation. The easiness, along with the modularity and flexibility of the system, facilitates the installation of small-scale solar projects. Moreover, an added advantage is that the installation can be expanded depending on the needs at a given time.

Multiple Purposes

Other than generating electricity, solar energy can also be used for different purposes. The heat can also be produced through solar energy. Solar energy can also be utilized to create power in regions where there is no requirement of an electricity grid. It can also be coordinated into the materials used for structures.

Most of all, the production of solar energy do not have any harmful effect on the environment. It also consumes very less water as compared to other energy production sources. It does not require any specific place; instead, it advantages rural and urban areas where electricity production is difficult or costs so much. The solar energy systems can easily be installed in both vertical and horizontal spaces.

Moreover, meeting the needs of solar power also costs less. It is much more affordable and economical source of energy when compared to other traditional forms of energy. These are the main reasons why many developed countries are finally adopting a lifestyle based on solar power.

Summary of the advantages of a solar system

Solar energy is most certainly the present and the future of the next generation. It is truly a safer, greener, and economical way of living life. It can be recharged anywhere and is a renewable source of energy, which causes no pollution in the atmosphere. Solar energy also puts fewer burdens on the other forms of non-renewable resources and protects the Earth from various types of pollutions. The electricity produced by solar energy systems can also be stored in rechargeable solar batteries. The development of solar energy and its technologies that are clean, inexhaustible, and affordable, provides us with huge benefits in the long-term.

Chapter 5

The Disadvantages Of solar energy

Though solar energy is renewable and domestic, at the same time it is can be deemed unaffordable and inefficient. Solar energy also has other disadvantages, such as latitude, which is one of the main factors in managing the efficiency of solar power.

Inefficient in some locations

Not all locations get the same amount of annual sunlight, with the efficiency of solar power dropping dramatically the farther you get from the equator. Thus, there is an increase in the dependency on batteries, space for installation, and limited capacity of solar technologies.

Inaccessible

The solar energy is an adequately renewable energy source only in the presence of sunlight. Thus, solar energy is accessible to us as long as we have sun; otherwise, it is of no use. Solar efficiency is also determined by the season in a country as the sun is tilted closer to your location in summer so you can generate more electricity than you need. Whereas in the winter, the sun is tilted farther from your area, so one can't make enough electricity to supply their needs.

The sun's abundant power offers a limited infinite source of energy; it means that the energy production depends upon the sun. Therefore, electricity cannot be generated during the night, so it requires you to either store excess energy which is made during the day, or try to connect to an alternate power source such as the local utility electricity grid. It also means that an individual has to pay more on top of the high cost of solar energy systems.

There are diverse applications of solar energy, but they are also harming the environment in many ways. The photovoltaic and thermal energy which are generated through solar energy, requires additional equipment (inverters) during installation to convert direct electricity (DC) to alternating electricity (AC) to be used on the power network. Their efficiency levels are relatively low (between 14%-25%) as compared to the efficiency levels of other renewable energy systems. For the installation of solar power systems, the land used can be massive and leaves no room for sharing the land for agriculture. Solar power affects land use when it comes to the production of materials needed to produce photovoltaics.

Presence of toxic metal

There are so many poisonous metals found in solar panels, which are as follows: cadmium and lead, which are extremely poisonous metals. Other hazardous and toxic materials used in the production of solar panels are gallium arsenide, sulfuric acid, hydrogen fluoride, nitric acid, 1,1,1-trichloroethane, hydrochloric acid, and acetone.

Unsafe Production

The production of solar energy is considered risky than the production of power from fossil fuels, in a sense that due to higher usage of solar energy systems, there is an excellent habitat loss, and the use of highly hazardous materials in the manufacturing process affects the environment.

Solar energy production is a man-made process, so if we apply the second law of thermodynamics, solar cells can never reach 100% efficiency; though, the highest efficiency of such a system is 85%, and that can only be achieved through mirrors and motors to follow the sun. So, when there are systems that do not track the sun, the highest efficiency reduces to only 55%. Thus, tracking the sun in cloudy days is also a factor that shows a disadvantage of solar energy systems. The large solar energy systems can affect the environment near their locations. There is a lot of space needed, so clearing land for construction and the placement of the solar power plant may have long-term effects on the habitats of the particular area and nearby areas.

Though the solar energy systems do not produce air pollution, water pollution, or greenhouse gases, there are some toxic materials and chemicals which are being used to make the photovoltaic (PV) cells that convert sunlight into electricity. So, if solar thermal systems use potentially hazardous fluids to transfer heat, the leakage of these materials could be harmful to the biotic and abiotic environment.

All manufacturing processes require some water. Also, water is used to make solar panels, so if large volumes of groundwater or surface water will be used in some arid locations, it will affect the ecosystems that depend on these water resources.

Significant initial investment

The solar energy technologies usually require a significant initial investment, and an average-sized solar energy system can acquire a high cost, which can produce 4kwh to 8kwh. Also, there is an increase in technological improvements, expanding markets, and government subsidies and incentives, which takes much financial resources.

Labor Intensive

Solar energy, in general, is considered more labor-intensive than the other energy production sectors, and there are a lot of health issues with the workers of solar energy protection as the manufacturing of solar systems involves exposure to toxic metals.

Additional Cost

The solar power system does include a battery storage system so the power can be used in the night when there is no sun, which is an additional cost. The solar energy system can be connected to the electricity grid. Still, the battery storage systems are required if you plan on supplementing your energy needs by connecting to the local energy grid. Solar energy systems improve the security on the power grid against overloads or fires in transformer substations only if they are larger and occupy a larger surface area.

The local electricity loses some of the energy, around 3-5% during transportation and distribution. Still, the solar energy systems on the roof or in the yard of your homes don't even provide 85% efficiency.

Solar energy systems require different types of equipment for the installation, and it means that it has to be installed almost in both vertical and horizontal spaces. The easiness is only along with the installation of small-scale solar projects. For the larger-scale projects highly powered plants has to be used, which affects the habitat of the native animals and birds. Moreover, if you need to generate more power, you have to expand the area for the installation.

Summary of the disadvantages of a solar system

Though solar energy is most certainly the future of the next generation, it is somehow unsafe, as it is harmful to some inhabitants, such as insects, animals, and birds. It has to be recharged, and one has to replace different batteries for the proper working of solar energy systems. The electricity produced by solar energy systems can only be stored in rechargeable solar batteries, which do not have any guarantee. Solar energy is also unaffordable and inefficient in many ways. It provides us with benefits as well as long-term harm.

Chapter 6

Solar energy facts

Solar energy is the form of energy that we get from the radiation coming from the Sun. Scientists were trying to take advantage of solar energy as the other power generating natural resources like oil, natural gas, coal, etc. are becoming limited. Non-renewable resources are not enough to fulfill day by day, increasing individual requirements. First, scientists were trying to get that energy from the Sun, and now they are trying to save it as sunlight is only available in the daytime. In this chapter, we'll discuss some interesting facts of solar energy that force humans to shift from other natural resources to solar power.

Global warming

Solar energy is 100% renewable. We have zero worries about running out of solar energy like all other non-renewable resources. All human wants can easily be fulfilled by solar energy. According to NASA, they can nourish us with solar power for the next 6.5 billion years as well. The only problem is that the Sun is getting warmer day by day due to global warming. If this condition keeps on increasing, the Sun would burn the whole world.

Adequate supply of energy

It is the only form of energy that is adequate and abundant. The energy we get from solar radiation for only one hour is enough to power the house annually. The entire surface of the earth receives 120,000TW of solar energy. This energy is 20,000 times greater than the energy that the ground needed. It can provide power 24/7 and within seconds.

Free of cost

The solar energy is free of cost and the cheapest way to get power. No one has the right to claim the Sun as his property. Every part of the earth is enriched with this unlimited energy source in plenty. We have to install the solar panels in a large area to get maximum advantages of it.

Fastest way to get energy

Solar energy is the quickest way of getting power. Although the Sun is very far away from the Earth, i.e., about 95 million miles away, the energy rays only takes around 5 minutes to reach the land.

Solar cell (device to collect solar energy)

The first solar cell is built in 1954 by Bell Laboratories. Although solar energy is adequate to meet the whole world's needs, it needs an efficient way to become the main power source. A single solar cell produces one watt of electricity when it strikes with the light. This effect is also known as the photovoltaic effect (PV). PV was first discovered by Edmond Becquerel in

1839. Multiple solar cells, when are exposed to solar radiation continuously, generate thousands of watts of energy. These solar cells are expensive, but the profit they give is more than the expense. With time, the prices of solar panels have also fallen.

Pollution free

Noise and air pollutions are the biggest threat that the world is facing these days. One of the most interesting and unbelievable facts about solar energy is that it is pollution-free. While converting it into electricity, it does not cause a single type of pollution, whether it is noise pollution or air pollution. This source of energy is essential to have a pollution-free environment. Humans are making assumptions that shifting towards solar power will make the environment less polluted and even pollution free.

No fuel needed

Just like other non-renewable resources, solar energy is free of these kinds of supports. The only thing required to generate electricity from solar radiation is a photovoltaic cell that needs light. When the sunlight reaches the earth, it takes less than one minute to hit these cells. There is no need for fuel or any other thing to collect it.

Production of fossil fuel

Interestingly, the solar energy is used to produce fossil fuels as well. Sunlight is captured by plants, and it gets harvested in them. With the help of photosynthesis, solar energy gets converted into chemical energy. The chemical energy then creates biomass that ultimately forms fossil fuels. These fossil fuels are then utilized to gain power.

Solar energy can be stored

At night, there is no sun, and ultimately there is no solar energy. Does the problem arise of what should be done at night or in days when there is no sun (rainy days or cloudy days) to get advantages of solar energy? Solar energy is now storable as well. To keep it harvested for longer-term, modern technologies and devices like photovoltaic panels and solar thermal panels have been made. They have the capacity to store a large amount of energy for a long time.

Satellites and technologies

Photovoltaic cells are responsible for capturing solar radiation and for converting them into electric energy. At first, the PV cells used to power satellites and airplanes instead of houses. Vanguard 1 is the first satellite that is powered by solar energy. It was made by the US in 1958. This satellite weighs only 3.5 pounds and operates entirely on solar energy. The satellite is still in space and now considers the oldest satellite powered by solar power.

Another solar-powered technology is the first solar-powered bus, "Tindo." It entirely operates on solar power. The main benefits of this bus are that it is free of operational cost and pollution-free. Solar energy is not only used to power satellites and houses, but it is also used to power calculators and other domestic appliances.

Solar airplanes

Scientists have also formed solar planes. The successful results of this project prove that air traveling is also possible with the help of solar energy. Bertrand Piccard made the first solar-powered plane. This plane traveled the whole world without using another fuel or engine system.

Popular countries using solar energy

China is one of the most popular and common countries using solar energy. The world's most significant and unique solar projects like the Longyangxia dam and Sun grow solar farm are made and discovered by China. They are generating a massive amount of energy that is taking China's economy to the next level.

Japan is also becoming parallel with China. Japan is one the first country that uses solar power to generate energy. After the successful solar projects, every country is making an effort to get benefits from solar energy.

India is taking maximum advantages of it. Installing solar panels over the canals not only protects the water from evaporation but also capturing sunlight to get energy. This idea is benefited in two ways: protecting from loss of water and making electricity.

Australia is the country that is facing a high number of radiations. If it uses this energy, it can be the first country that is fully powered by solar energy. It is already a star in holding household powering by solar energy.

Water cycle's completion

The water cycle is generated due to solar insulation. The sun rays cause the temperature to rise, which results in the evaporation of water. The water vapors then go upward and form clouds due to condensation. The water then comes to the clouds in the form of rain and gets absorbed in the earth's surface or mix with oceans and lakes etc. In this way, the water cycle continues.

Other facts

Solar energy has many advantages but has only two or three drawbacks. The solar energy is measured in kilowatt-hour that is used for larger units (1KW consist of 1000W). Only 40% of the total solar energy gets into the earth, which consists of infra-red radiations, UV radiations, etc. We have excess to only 30-40% of the total solar energy as the remaining gets absorbed into the earth's surface. The rays that hit the ice reflected into the upper atmosphere. Only this limited amount of energy is enough to fulfill individual requirements for several years; then think what the total percentage can do.

There is no need to install solar planes in every house, and there are other numerous options like, a community solar service through which homes are connected to get electricity generated from solar power. The main benefit of using this electricity is that it is less expensive and is available every time whether the Sun peeps through your window or not.

The fact about the usage of solar energy that we should ponder upon is that the solar energy causes emission of greenhouse gases. These rays are somehow harmful. The main disadvantage of these rays is global warming.

Solar energy systems usually provide an ideal electrical power source for navigational beacons, satellites, outdoor lighting, and water pumps in remote areas.

Solar energy can be used to run a turbine by heating the fluid of it to produce steam. This steam spins a turbine that is connected to an electrical generator. This is done through concentrated solar power systems that use mirrors to reflect and concentrate sunlight onto a small area.

Solar energy is the cheapest and over-sufficient form of energy having limited side effects. The use of this energy can take the world to the next level. Many plans are possible only with this energy that cannot be achieved by other resources. Non-renewable resources are getting limited day by day. The only possible way to overcome them or to substitute them is with solar energy. In the near future, the use of solar energy systems will likely continue to increase in those areas where sunshine is plentiful. The solar energy system will most probably be placed on rooftops, in yards, over parking lots and along highways, and then will be connected to electric utility's power-line system so a lot of money can be saved.

Chapter 7

The basics of electricity

If we want to learn the basics of electricity, then our primary focus should be on the following:

- An individual should be well known of conductors, semiconductors, and insulators, their explanations, and the main differences between them.
- The information of current and how it flows.
- The laws which are used in electricity like; Ohm's Law to calculate voltage, current, and resistance.
- As electricity cannot flow in a continuous form in any circuit, one should know about resistance theory and how to apply it in electricity production.
- The voltage maintains the electricity because everything running on electricity needs a specific voltage, so one needs to calculate the voltage drop across a resistor.
- The calculation of the power of electricity is given with other fundamental values.

- Along with all those maintainers of electricity, the other things to identify are the factors that determine the polarity and strength of a current-carrying coil's magnetic field.
- Also, determine peak values, instantaneous values, and effective values of an AC sine wave. While studying inductive reactance and capacitive reactance, one should also identify factors that affect parallel circuits, inductive reactance, and also capacitive reactance in an AC circuit.
- On the basics of electricity, the calculation of total impedance of an AC circuit along with the explanation and difference between real power and apparent power in an AC circuit.
- The calculation of the kVA of a transformer is also essential.

Conductors, Semiconductors, and Insulators

There is an electric current that is produced when free electrons move from one atom to the next. The materials that allow many electrons to move freely are called conductors. Copper, aluminum, silver, zinc, iron, and brass are considered good conductors of electricity. Copper is said to be as one of the most common metals used for conductors and is relatively sharp.

The semiconductor materials, such as silicon, can be used to manufacture devices, such as those that have characteristics of both conductors and insulators. When an external force is applied from one direction on semiconductor devices, those devices will act as a conductor. Oppositely, when the external force is used in the opposite direction, the semiconductor devices will serve as an insulator. Now, the materials that only allow a few electrons to move freely are called insulators. Materials such as plastic, glass, rubber, ceramic, and mica are good insulators. Transistors, diodes, and other solid-state electronic devices run on this principle of semiconductor materials.

If we take into account an example of an electric cable, we can easily understand how conductors and insulators are used. Since, copper is a good conductor, electrons flow along a copper conductor to provide energy to any electric device such as a lamp, radio, or a motor. An insulator, which can be plastic, rubber, or glass around the outside of the copper conductor is then provided to keep electrons in the conductor.

The phenomenon of electromagnetism is used in conductors, which are explained as follows:

Electromagnetism is a magnetic field generated by the flow of current in a conductor. Every electric current produces a magnetic field. A definite relationship surely exists between the direction of the magnetic field and the direction of the current flow. The left-hand rule for conductors demonstrates the direction of the magnetic field in the current relationship. If a conductor that is carrying current is grasped with the left hand with the thumb as pointing in the direction of electron flow, the fingers will point in the direction of the magnetic lines of flux. The coil which carries the current is of wire carrying a current, and it acts like a magnet as it is a wire of reel, so its loops act as small magnets. The individual fields of the wire sum up together to form one magnet. The strength of this field can be increased by adding turns to the coil. The strength of electromagnetism can also be increased by increasing the current. The electromagnet in the process of electromagnetism is composed of a coil of wire wound around a core. The core conducts magnetic lines of force with relative ease and is usually a soft iron. When current attracts the charges, so when it is passed through the coil, the core becomes magnetized. The electromagnets have the ability to control the strength and direction of the magnetic force. The opposite poles of a permanent magnet always attract each other. An electromagnet is made to control the strength of its magnetic field, which then controls the strength of the magnetic poles. Thus, a large variety of electrical devices such as contactors, motors, relays, circuit breakers, and motor starters use electromagnetic principles.

- Current

Electricity is produced through the flow of free electrons in a conductor from one atom to the next atom in the same universal direction. This flow of electrons is known as current and is designated by the symbol "I." As we know that the electrons move freely through a conductor at different rates and electric current has different values, the number of electrons which passes through a total cross-section of a conductor in one second determines the flow of current. It takes almost 1,000,000,000,000,000,000,000,000 total atoms to fill one cubic centimeter of a copper conductor. It would be tough; writing 24 zeros after the number 1, write 1024. If we even try to measure small values of current, it would definitely result in unimaginably vast amounts. That is why the current is measured in amperes, which is abbreviated "amps." The letter "A" is used as the symbol for amps. One amp of current means that in one second, about 6.24 x 1018 electrons move through a cross-section of the conductor.

The direction of current flow

One more aspect to be understood about current is the direction of current flow. There is a difference between electron flow and current flow. Conventional contemporary flow theory states that current flows from positive to negative. In this book, we will use the electron flow concept so that we can clearly understand the difference between electron flow and current flow.

After studying current, another vital thing in the basics of electricity is voltage, which is explained as follows:

- Voltage

Voltage is the force which is applied to a conductor that causes an electric current to flow. Let's take an example of water flowing through a pipe and compare it with the production of electricity. A force that might come from a water pump or gravity is required to get water to flow through a pipe.

Similarly, as water flows through a pipe, the current flows through a conductor. Electrons are always negative and are attracted by positive charges. The electron will always be attracted from a source having an excess of electrons, thus having a negative charge to a source having a deficiency of electrons, which has a positive charge. Voltage is shown by the letter "V," or the letter "E." The unit of measurement for voltage is volts.

Now we will focus upon the Voltage Sources. An electrical voltage can be generated in various ways. A car's alternator utilizes magnetic induction; a battery uses an electrochemical process, and a power plant generator also employs a magnetic induction process. In all of the voltage sources, there will always be the characteristic of an excess of electrons. These excesses will be at one terminal and a shortage at the other terminal. An electrical drawing by two lines symbolizes the terminals of a battery. The longer line indicates the positive terminal. The shorter line indicates the negative terminal.

Units of Measurement

Prefixes that are used for values of voltage: 1 kilovolt = 1 kV, 1000 V = 1 millivolt 1 mV = 1/1000 V, 1 microvolt = 1 mV 1/1,000,000 V.

- Resistance

Another factor that plays a role in an electrical circuit is resistance. All the materials used to hold back the flow of electrical current to some extent. The amount of resistance depends upon many factors such as:

- Composition
- Length
- Temperature
- A cross-section of the resistive material

The rule of thumb for resistance is that the resistance of a conductor increases with an increase of length and a decrease of the cross-section. Resistance is shown as the symbol "R." Its unit of measurement is ohms (Ω).

Resistance Circuit Symbols

Symbolically, resistance is usually indicated on an electrical drawing in one of two ways. An unfilled rectangle is commonly used along with a diagonal line. Resistance can be in the form of various components. For example, there might be a resistor placed in the circuit, or the circuit may contain other devices that have resistance.

Prefixes that are used for the values of resistance: 1 kilo-ohm 1 kΩ = 1000 Ω, 1 mega-ohm 1 = MΩ 1,000,000 Ω vice versa.

- Ohm's Law

The relationship between current, voltage, and resistance was studied by the 19th-century German mathematician, George Simon Ohm. Therefore, it is named as Ohm's law. Ohm's law states that the current varies directly with voltage and inversely with resistance. It is considered mandatory for electrical designers to decide how much voltage is needed for a given load, such as computers, clocks, lamps, and motors, as every device needs a specific amount of voltage. That decision must be made by putting in mind the relationship between current, voltage, and resistance. All the analysis and the electrical design begins with Ohm's Law.

Ohm's Law triangle

An easy way of making a decision is Ohm's Law Triangle. By arranging voltage, current, and resistance in a triangle, one can quickly determine the correct formula.

Example:

An example of applying ohm's law is as follows: assume that the Voltage Ohm's Law supplied by the battery is 20 volts, and the resistance is 10 Ω. Mask the "I" in the triangle and use the resulting equation. I = I

= 2 Amps E R I = 20 Volts 10 Ω Using the same circuit, assume the ammeter reads 300 mA, and the resistance is known to be 20 Ω. To solve for voltage, take the "E" in the triangle and use the equation which comes in its result. E = I x R E = 0.3 x 20 E = 6 Volts. Always use the correct decimal equivalent when dealing with numbers that lead to micro (μ), Milli (m), or kilo (k).

- Parallel Circuits

In this book, we will be discussing three different types of parallel circuits that are DC series circuit, DC parallel circuit, and the series-parallel circuit. For DC series circuit, a resistance, voltage, and flow of current. A series circuit is formed when several resistors are connected end-to-end leading to only one path for current to flow in a circuit. The resistors can be original, or it can also include the devices that have resistance. The values of resistance sum up in a series circuit. The value of the flow of current in a circuit can be calculated by using Ohm's Law. The voltage has to be kept in a specific amount in a series circuit, and the voltage can be measured across each of the resistors in a circuit. The voltage across a resistor is referred to as a voltage drop. The DC parallel circuits are formed through proper resistance, voltage, and flow of current. When two or more resistances or parallels are placed in a circuit side-by-side so that current can flow through more than one path. Mainly, there are two paths of current flow in a parallel circuit. One path is from the negative terminal of the battery through a resistor R1 returning to the positive terminal. Then, the second path is from the negative terminal of the battery through resistor R2 returning to the positive terminal of the battery. The accurate voltage is given when the resistor is placed in parallel across a voltage source. The voltage will be the same across each resistor. Each resistor has 12 volts available to it. The flow of current through a parallel circuit divides and flows through each branch of the circuit. The total current in a parallel circuit is equal to the sum of the current in each branch of the circuit. The current flow in a parallel circuit is equal when equal resistances are placed in a parallel circuit. The current flow is in opposition to value resistors is the same in each branch. Whereas, the series-parallel circuit is also known as compound circuits. To build the series-parallel circuit, at least three resistors are required. This is the same as DC parallel circuit.

- Power

Whenever there is any motion due to a force of any kind, work is accomplished. For example, work is done when a mechanical force is used to lift a weight. If a force were exerted and it doesn't cause any motion, then no work is done.

When voltage is applied to a conductor and electric power is generated in an electrical circuit that will cause electrons to flow, it shows that the voltage is the force, and electron flow is the motion. The rate at which the work is done is called power and is represented by the symbol "P."

Unit for measurement of power

The unit for measurement of power in watts and is represented by the symbol "W." The rate of work is done in a circuit when 1 amp flows with 1 volt applied. It is known as a watt. A power consumed in a resistor only depends on the amount of current that passes through the resistor for a given voltage. It is expressed as voltage times current - P = E x I or P = EI.

Heat

Whenever the current flows through a resistive material, it causes heat. If the temperature is too high, then an electrical component can be damaged. That is why electrical equipment is often rated for maximum wattage. The more heat the equipment can dissipate when the wattage rating is higher.

Mostly, it happens that the supply of current for electrical devices comes from a direct current source (DC), but sometimes we need an alternating current source (AC). In an alternating current source, an AC generator is used to make electrons flow first in one direction then in another. Another name for an AC generator is an alternator. Terminal polarity gets reversed in the AC generator many times a second. Through a conductor, the flow of electron will be from the negative terminal to the positive terminal, first in one direction then another. There are two phases through which the power is supplied in the AC circuit – a single-phase AC power and a three-phase AC power. Single-phase power is used for small electrical demands such as the electrical appliances which are used in the home. Three-phase power is used where large blocks of power are required, such as industrial areas, commercial applications, and moreover for industrial plants.

Chapter 8

How solar panels work

The following are the two types of technologies that give solar energy.

- First, solar photovoltaic that uses semiconductors to convert sunlight to electricity.
- Second, solar thermal produces electricity by heat energy from the Sun.

How this system really works?

Just like a wind turbine or power plants, solar panels don't produce any energy by moving its parts. Then how exactly, power is generated from the solar panels. Solar panels consist of conductors and silicon. These silicon's PV cells are protected by a durable glass protector, another layer lies under the glass protector is insulation sheet that saves the panel from high thermal temperature and internal humidity, This insulation sheet is necessary for efficient working of a solar system, as extreme hot and cold weather can slow down the working of the solar pane. Anti-reflective coated surface intakes all the sunlight spectrum to provide the silicon cells with more and more radiations for absorption.

Silicon cells are of two (2) categories -

Monocrystalline and polycrystalline

Monocrystalline composed of only one silicon crystal that provides more space for electrons for movement resulting in high efficiency as compared to polycrystalline; thus, are very costly. On the other hand, polycrystalline are composed of multiple silicon crystals but are considered efficiently lower than monocrystalline.

Another form of solar cells are thin filmed cells that have silicon cells coating on the surface of the glass to create silicon sprayed module instead of introducing individuals cells. They are lower in cost and have lower working ability.

When the sunlight heat energy falls on the panel, the heat energy from the Sun changes the natural state of electrons that are present in the silicon by vibrating the electrons, meanwhile the solar panel is connected to an electric circuit.

Basically, in a solar panel, photon eliminates an electron from their atoms that produces electricity known as the photoelectric effect, an electric current is produced when these released electrons are trapped, that current later utilizes as power electricity. Photons are considered as a packet of light.

History of photoelectricity

Back in 1839, a French physicist Edmund Becquerel decided to focus on one to focus on the photoelectric effect. He was the one to notice that whenever light falls on certain things, current is created as a result.

Every solar panel consists of photovoltaic cells PV that is sandwiched between semiconductors such as silicon. These semiconductors converts' sunlight heat energy or photons into electricity.

Semiconductors can be in different forms. Whether crystalline cells, flat or thin layer of cells, these cells connected with the help of wire and electric connectors are sealed together and are weather-resistant.

These semiconductors are doped with different matter and turn into positively and negatively charged particles.

PN Junction

Phosphorus is usually used as a doping agent and named as N-type to give the top layer of a negative charge. Boron is commonly used for giving a positive charge of the bottom layer, which is named as P-type. The whole process of P and N-type silicon is known as PN junction.

One side of semiconductors is positive, while other is negative. For this purpose, a huge area is required for the consumption of heat energy from the Sun. Photovoltaic power plants have the ability to provide power electricity to a large number of houses. It's possible to light up the whole house if the patterns of solar cells in photovoltaic panels and patterns of many photovoltaic panels are placed in an arranged form.

When sunlight falls on the panel, the sun energy strikes straightly to the solar cells. This separates the electrons from the atoms. A circuit is produced when conductors are connected to both positive and negative ends, and the free electrons are taped as an electric current known as electricity; current varies depending upon the quantity of light falls on the module.

When various number of solar cells are attached with each other in a proper arrangement results in a photovoltaic module, these modules provide about 12 volts of current. Large amount of power is produced when various numbers of modules are attached with wires in the form of a series array to provide high DC volts. But if various numbers of modules are linked in parallel form, then one inverter is enough to work.

Multi-junction and single-junction devices

Single junction photovoltaic material is commonly in use. But in a single junction there is an equal band gap between each cell layer. Photons that have decreased energy is less useable because these lower-energy photons are not able to free an electron from an atom, whereas high energy photons from the sunlight are required that can release an electron in an equal bandgap to create an electric circuit.

On the other hand, a multi-junction consists of more than a single junction that produces current by changing the thermal light spectrum into electricity. Its vertical stand of single-junction has a bandgap in descending order - thermal photons with increased energy are absorbed by the cells that are located on the top and transfer the decreased energy photon to the bottom cells. The PV cells in an array form generate direct current, but this direct electricity is not yet able to give power to your house. Other materials are also required for this purpose. At accurate frequency and voltage, it's important to transform the direct current output into the AC. An electric inverter fulfills this purpose.

Inverters change the solar energy to usable electrical energy, and inverters transform the direct current of solar panel into the alternating electricity. This process help to provide usable electricity to our homes.

This usable electricity from the solar panel gives electricity to the home appliances; it's just similar to the current from the electric circuit. Sometimes solar panels are not enough to give electric power to the whole house. There is also a solution to this problem. Advance electric grid produces more electricity for your use whenever you need because advance grid connection modules generate about 40 direct current voltage and consist of about 48 to 72 cells.

When we use less electricity, the grid stores the leftover solar electricity for further use. So, it's beneficial to have a power grid alongside the solar system which enables us to utilize electric power in excess, depending upon the electrical need.

Solar panel performs its given tasks only in the daylight. It only produces solar electricity in the sunlight; but what about solar electricity in the night? We also need power in the nighttime for use. In the nighttime, power grid plays its important role in providing stored power for the nighttime use.

Does cloudy weather affect the working of the solar system?

Solar panels work when they are exposed to light; that means solar panels can also work in less light or cloudy days. In short, the stronger the sunlight, the stronger the electricity.

Working of solar panels in summers

Summers are the best for solar system because it produces more energy in warm weather. In this era of solar advancement, the system has expanded in the U.S. and other countries. The use of solar panels within the United States sees the use of a national grid that intakes power. The quantity of electricity used, and electricity is stored in a national grid must be in equilibrium. These PV grids cover up 99% of the global solar system market.

Is it possible to measure solar power electricity or not?

Have you ever heard about net meter? Net metering provides a free storage solution during night or bad weather forecast. The net meter device measures the amount electricity produced by the solar system. When you fix a solar panel in your home on the rooftop, it is linked with the home utility meter. This meter is placed outside the house and is similar to electric meter. The power that is produced in excess goes back to the grid for storage replacing energy credits, during hot day, but these energy credits are utilized in night when the solar system is not exposed to light. In short, it saves money with its electricity storage.

Another interesting thing is that the solar system never runs out of electrons. It's impossible for a solar system. The reason is that the photons release electron in the doped semiconductors, which generates the electric circuit and, again, that released electrons go back into the semiconductors of the solar panel.

New advancements in solar technologies have introduced portable solar panels. The portable solar system is easy to transport as compared to heavy and fixed traditional solar panels. With the passage of time, solar systems are becoming more and more common as they are widely used these days, and they are easily available; it's the best way to save yourself from spending a lot of money on electric bills.

Chapter 9

Facts about the solar panel

Solar energy has so many advantages that other resources like fossil fuels hardly have. Environmentally friendly energy having only one or two disadvantages is such a blessing. Non-renewable resources are unable to meet the worldly needs. They are at their end-stage. This situation forced the human to make solar panels. Solar panels catch sunlight and use it to generate electricity. They are of two types:

- Photovoltaic panels
- Solar thermal panels

The first one converts solar light into electrical energy, and the second one is responsible for heating water and absorbing heat from the Sun. Now it is fast-developing energy not only in Australia but also around the world, as its need is increasing day by day. Here are some interesting facts about solar panels that will blow your mind.

First solar cell

No one can deny the importance and benefits of solar energy. There is not a single day in which solar energy does not reach the ground. Solar energy reaching earth for one hour can fulfill the energy requirements for up to one year. Leonardo Da Vinci was the first that predict that there would be a method or technology that will use sunlight or sun power to heat the water or to produce electricity in 1447. Greeks and Romans first used the passive design of solar panels. They made their houses in a way that the windows of the houses are towards the South. That allows the sunlight to enter the house and kept them warm.

The ability to produce electric current from the sun rays is known as the photovoltaic effect. This effect was first discovered by a French physicist Alexandre Edmond Becquerel in 1839. Solar cells are made to capture and for harnessing solar rays that have a photovoltaic effect. The first solar cell that has silicon coating was made in 1954. Bell Laboratories have been honored to have this opportunity. At that time, the efficiency of this cell is only 1%. Later, efficiency increased, and now the solar cells have efficiency up to 40%.

Production of heat without facing direct sunlight

As we know, the Sun only appears in the daytime. The sunlight facing earth is greater in hot days as compared to cloudy days. At night, there is no sign of Sun or sunlight. Hence, there would be no concept of solar energy. But as the human believes that nothing is impossible, he has found a solution to this problem as well. This problem is solved through a storage system. Solar cells are designed to catch some spectrums of sunlight, not the whole ray. This designing idea helps in the working of solar panels, even on cloudy days. Remember, the hues that are coming from the sky are also used by the solar panels to generate electricity.

Environmentally friendly

Solar panels are designed and manufactured in a way that they produce no pollution - no air pollution, no fuel burning, no noise pollution, etc. Interestingly, there is no waste of material produced by it. It reduced the emission of not only carbon dioxide, but also of other gases. Monthly, using solar panels to produced energy helps in reducing sulphur dioxide up to 8 pounds, nitrogen dioxide up to 5 pounds, and approximately 1400 pounds of carbon dioxide. Just imagine, one-month usage of solar panels can reduce this much of pollution, wouldn't its permanent usage bring much environmental change?

Production of energy

For solar energy, the kilowatt per hour unit is used. 1-kilo watt is equal to 1000 watt. This unit is used for larger quantities. 1-kilo watt of energy can power a 100-watt bulb for almost 10 hours. It is also equivalent to the energy that is produced by burning 170 pounds of coal and 300 pounds of carbon dioxide. Nowadays' solar panels have a capacity of 225 watts, i.e., they can produce 20 watts per square foot. As per calculations, one house needs approximately 30 solar panels for running. Just think for a while, per square panels are generating this much of energy, then what amount of energy can be produced by the panels that are spreading over a vast area of land?

Community effort

It is not permissible for the public to install solar panels individually on every single house; now it becomes a community effort. The solar panels are being installed on a large scale to produce a large amount of energy that share with every household. It is only allowed for the buildings or factories that need more than 20mm of energy produced by solar panels.

Perfect area for investment

It is an ever-growing business. It does cause not only an increase in the economy but also causes a decrease in bills and other things. In 1977, the price of one kilowatt of energy was approximately $76.67, and it is decreased up to $0.30. More importantly, the installation of solar power panels is very easy as compared to other resources. They can be placed in mobile and calculators and can be installed over large buildings' roofs in a large number. The more power plants, the more solar energy, and the more increase in saving. Some countries have installed so many plants that they are now selling their power plants and are getting unlimited benefits. Also, solar industries are providing more jobs for people than the jobs that are normally available in the job market.

Solar power installation

Solar power installation is rapidly increasing business. The leader country in solar power installation is Germany. Italy, Japan, and the USA are also on the top of the list. The USA is known to be the third-largest country in installing solar panels. Day by day, its price is decreasing. Increasingly it is appealing to other countries. Many Americans have shifted towards this form of energy. US solar power installation is increased 23 times than the number of solar panels that were installed eight years ago. Countries are taking it as challenged and are installing solar power stations in a unique way that gives them more than one benefits. For example, there are solar farms in China. These plants are installed on floating destroyed coal mines. This gives solar energy as well as protects solar panels from overheating, which ultimately increased their efficiency. Canal solar power project in India is another example of this. It gives both solar energy and protection from water loss due to evaporation. California has the largest solar power plant built in the Mojave Desert. This plant is producing 90 percent of the total solar energy that is being produced in the whole world.

Versatility and maintenance

Solar panels are designed in every shape and size and fit on every type of roof, buildings, etc. They can power not only phones, calculators, houses, etc., but also airplanes, satellites, and even in the future, cars and boats. Large and empty areas provide the best locations for solar power installation, as they face direct sunlight and can installed large and many solar panels.

Solar panels require no installation at all. Once they are installed, there are no taxes or any other bills. For better performance, it is better to made roofs in a way that the roof faces south at an angle of 30 degrees. It provides space for sunlight exposure. Interestingly, the performance of the solar panel decreases in hot weather. They just required protection from wind and rain as continuous exposure of rain and wind can harm them. Now, solar panels have 25 years warranty and are much more efficient than the previous models.

Solar panels are subjected to give an enormous amount of benefits. No one can ever deny the importance of the Sun. Due to several benefits, the Sun is ruling over the entire world. These solar panels not only work in daytime but also give an advantage at nighttime, as they have installed storage systems. Secondly, the low maintenance and its easy installation make the world more comfortable. Most of all, there is no danger of its extinction. It is 100% renewable, and every day we enjoy its benefits. Now, it is very easy to take advantage of this. Just call or email to the nearest solar office and enjoy this service within days.

Chapter 10

Buying a solar power system

Energy is an important facet of advancement. But the use of traditional sources of energy like petroleum and coal is now scarcely available. This is the reason there is an increase in the need to switch to a better and innovative energy source. What better way to gain energy from any other human-made sources when the natural and most primary source is right at the top. A solar power system is effective, and a must-have technology in today's era, where there is a shortage of electricity everywhere in the world. Why waste so much money on paying these utility bills and going through the ruckus of electricity shortage when you can directly get it from nature. That's why this system of solar panels invented by the great minds of the 19th century has helped humankind in numerous ways possible.

However, just like any other technology, this invention has also lead to people gaining profits by advertising and selling their products to create a profitable business. Therefore, it comes down to the customers on how wisely and effectively they choose when purchasing a solar panel system for their homes. Buying a solar panel system is a huge task and one that would stay with you for a long time, so don't get carried away with fancy advertisings and focus more on all aspects of the system before buying a reliable solar panel.

Following are a few tips for buying a solar PV system:

1. Taking suggestions and advice

Firstly, before even going to the shop, ask the people around in your neighborhood or extended family, who have these systems installed, about everything they know about the solar power system. Ask them about the brands they have chosen and take suggestions on what other companies are selling well in the market. You'll get to know of their experiences and might know any difficulties that they faced using the system. In this way, you'll be more able to buy an affective system because experiences matter a lot.

2. Warranty

The most important and crucial thing to check for is the warranty of the system. Make sure it is as long as at least 25 years, because if you are investing this much, then it should stay with you for a long time. Also, make sure that the manufacturing company is legal, otherwise the warranty is of no use. The company you make contact with must provide you with desired support and service. Therefore, deal with a trustworthy company who is willing to give a solar panel replacement in case of defaults.

3. Price check

Now the thing in buying a solar power system is that you can't expect to have a good, reliable, long-lasting, and effective system at a low price. Anything that is of quality value like the equipment and installation, isn't easily available at a low price. Also, you get what you pay for, so you have to pay enough amount for this one. As there are companies to sell their low-quality equipment and sell it at a low price, so it's better to pay more for a high-quality system rather than regret the choice you made on a low-quality system. Because the benefit of investing in a high-quality system is that you later have to pay only the minimal maintenance charges, and the system stays with you for a long time.

4. Certified

Another important step before buying a solar panel system is to make sure the government certifies it and that the company is reliable and honest in what it claims its products to be. Avoid anything that looks remotely illegal in the whole process of purchasing the system. Also, the system should be tested in a laboratory before putting it in the market, so this certification is a necessary thing to look for. Therefore ,look for the certification of your system that includes the testing done on the system.

5. Type of panels

Day by day, there is ongoing advancement in the manufacturing of the panel system with the inventions of monocrystalline panel technology, which can absorb a large amount of solar energy. Then there is polycrystalline, which is less efficient but cheap; and few other thin-film technologies. So, three things you need to consider are the type of panel, the roof size, and the output you need. The amount of panel should be adequate, not more not less, and one that suits your roof size. Moreover, there are specific types of panels that are suitable for homes:

1. Grid-tied residential solar power system with batteries are costly and difficult to install, but the power output is high.

2. Grid-tied residential solar power system without batteries is cheap, easy to install as it needs only a few components to be assembled.

3. Off-grid residential solar power systems are installed in areas where services are not adequate, so this system involves full energy independence.

6. Mounting check

The area that you live in also affects the purchasing of the type of panels. For instance, if you live in a cyclone-prone area, then make sure that the solar panel mounting is also cyclone rated. It's an important thing to check for that (which most companies don't pay heed to). Check if the system can withstand the changing weather conditions. In any cyclone-prone area, the mounting should be done on the roof or the floor for securing the system. Also, if you are living in a sunny area where there is sunlight most of the year, then monocrystalline panels are preferable. If sunlight is scarce in your area, then it's better to opt for polycrystalline technology, and make sure that where you place your solar power system, there is no obstruction between the Sun and the panels.

7. Inverter efficiency

The inverter is the main component of the panel system that converts DC into AC that can be used in our homes. Therefore, the inverter needs to be of high quality and efficient, so that less electricity is wasted in the conversion, and it would be beneficial for you, as it will directly impact your savings.

8. Installation process

The whole installation of the solar panel system is a task that requires professional handling and should be done by experts so that the system works efficiently and effectively.

9. Taking quotes

It is important to look for a few sellers who are trending in the market and ask them first to send you a few quotes of the panel system, which may include the prices of the inverter, the panels, and the whole installation process; so that it may help you in finding the right quote or package that suits your budget.

10. Taking lease or buy

Now, as installing a solar panel system is although very effective, it requires a good amount to be paid for it that not everyone can afford it; therefore, people sometimes take the system on lease. It's a good measure, but one thing to consider here, is that buying a system is beneficial in the long run, whereas leasing lets you access cheap electricity with not enough benefits. Make sure you compare the benefits and the cost of both the lease and buy before choosing the one that fits you the best. Remember that the lifespan of the inverter is shorter than the whole panel, and you have to replace them all once in a while, so make sure to count the cost if it in your budget too.

11. Contract details

The contract you sign before buying the solar panel system from any company should include all the important details like finance, ownership, warranty, and performance expectations. Make sure everything is in there before signing the contract.

12. Tackling sales pressure

One thing to consider if you want to have a good and effective solar panel system, is not to get swayed by the pressure of the salespeople and get impressed by their fancy advertising tactics. Buying a panel system is a task that requires effort and time, so an on the spot decision often turns out to be less beneficial. Before making any decision, ask the salesperson about the quality of the service, how easy the installation will be, and what problems might occur in the process. If you get satisfactory answers to these three things, then you should proceed with your buying.

These were the few important and necessary tips to keep in mind at all costs before buying any solar power system for your household.

Chapter 11

Avoiding common mistakes

The demand for energy is increasing rapidly. In the same way, non-renewable energy resources like coal, oil, and natural gas, etc. are at their end-stage. Human are intensely in search of energy that should be renewable, requires less input but give more output. It is looking insane for a while, but there is an energy resource that fits all the desires. That incredible energy is known as solar energy. For taking maximum advantages of solar energy, solar panels consists of a photovoltaic cell. These cells generate electrical energy when light strikes them.

Many people face a problem while using solar panels. Although it has unlimited benefits, it creates problems due to a lack of knowledge. Here in this chapter, we will discuss mistakes that should be avoided while buying or installing solar panels.

Misunderstanding about installation on roof

It becomes a myth that the roof must face north direction. It is not 100% true. If your roof is facing other then north direction, it's okay. It will decrease the efficiency of the solar panel but not that much. Keep in mind that the solar panel should be protected from wind, extreme sunlight, and rain, etc. It should not be shaded so much that panels cannot catch enough sunlight to generate power. The solar panels installation should be according to the structure of the roof.

Not have multiple opinion

While doing any work, we take suggestions from the people. Taking suggestions is mandatory as it helps to make a better decision. The same is the case with buying solar panels. It is better to take advice from the people that are using this technology or have knowledge about this. Visit companies that are well-known or have a better reputation. There are a lot of famous companies out there that have a bad reputation. These types of companies charge heavy amounts and in return, give fake panels. These panels will get destroyed within years. Beware of those companies. Decide on clearing all the doubts and knowing all the sides.

Cheap solar panels

Buying cheap solar panels will give you lifetime tension. Losers cannot be equal to winners. Doing so to save money will cost more than you think. Because they will destroy or give poor work after one year or less. Then you have to replace them or repair them, and ultimately you have to bear a loss. It is better to buy once, but of better quality. It is better to take advice from a professional solar installer. The professionals will also help you in choosing the perfect numbers of solar panels. What type of solar panels will be needed to run the entire building? A good solar panel might cost you too much at the beginning, but it will give you longer relief.

Not paying attention to warranty conditions

Always buy a solar panel that has a longer warranty span. Buying solar panels having no warranty or having less warranty is not the right decision. Always pay attention to the solar panel's material and performance warranty, inverter warranty, and installation warranty. Normally the best solar panels have a warranty of up to 25 years. This span is considered a standard warranty span. Try not to buy panels having less than 25 years warranty.

Inverters are part of solar panels. They can convert solar energy into electrical energy. The standard warranty life span of invertors is almost five years. Many others also have a warranty for more than five years.

Clean energy council recognition

The Clean Energy Council is an organization in Australia that keep notice of all the affairs. It ensures that safety standards and other basics guidelines are there or not. If the Clean Energy Council does not recognize your solar systems, then the government is not responsible for rebates. Also, it will cost you more then you expected.

Not having safety measures for the future

Like all other things, it is necessary to have a backup. As there is no sunlight at day time or during cloudy days, it is better to have a battery storage system. For the storage of solar energy, you need to install an extra solar panel. A professional will help you to know if there is any need for molding the panels, or there is space to install more panels for storing. Do not take the final decision immediately. Do it when you think it is okay, not when you want.

Not having knowledge of consumer's right

When you are purchasing something, you have some rights. Many people do not know about their rights, and due to this, they have not got many services that they should get. You should have proper guidance and awareness of your needs. It is also the responsibility of the salesperson to give you knowledge about the solar panel's reliability, investment procedure, and duration, warranty life span, materials used in the manufacturing of the panels, etc.

Doing it yourself

It is a very common mistake that most people do. Everything needs knowledge as well as practice. For every work, there are professionals out there that have more grip on this field that the non-professionals have. If you are un-professional or thinking of installing solar panels by yourself, then do not do it. Any mistake will cause a problem for you. The problem, in return, will cost you money as well. The money you want to save by installing panels yourself may result in a loss of more than the savings.

A professional knows what, where, when, and how panels should be installed. Let them do their work; they have more understanding than you.

Having useless expectations

There is a famous quote stating that expectations can kill you. It is very much true. Solar energy has double the number of pros than cons. Having a question in mind that is contemplating the worth of solar energy is a silly question. Waiting for new technology is also not worthy. The latest technology costs a lot of prices as well. The longer you wait, the more you pay.

Not taking care of solar system

Everything needs proper care. Like this, solar panels need proper care as well. Although they do not need maintenance, one has to work for its care and protection. They should give their maximum efficiency. Ignoring shadows may lower the efficiency of solar panels.

Hence, by avoiding all the mistakes and keep in view the important guidelines will give you benefits. If you will not pay attention, it leads to harmful consequences. The solar panel generates electrical energy, and electrical energy needs too much care. It can take once life if it is not carried properly.

Chapter 12

Batteries

Batteries are made up of two or more voltaic cells. Those cells are connected in series to provide a steady DC voltage at the battery's output terminals. A chemical reaction inside the cell produces the battery's output terminals voltage. Batteries work because of an electrolyte process. The electrodes are immersed in an electrolyte, which forces the electric charge to separate in the form of ions and free electrons.

Working of voltaic cell

The working of the voltaic cell is that it converts chemical energy into electric energy. Moreover, a voltaic cell consists of two different metal electrodes, an acid or a base that is immersed in an electrolyte. As a result of the chemical reaction, there is an immersion that produces a separation of charges. It produces current and current capacity increases with large electrode sizes. The anode of the cell is the negative terminal because it forms positive ions in the electrolyte. The opposite terminal of the cell is known as its cathode. The motion of electrons of voltaic cells in ionic bonding can be used to generate an electric current. A battery's current and voltage output ratings are determined by the elements such as an acid and a base used for the electrodes. It depends upon the size of the electrodes and the type of electrolyte used in a battery.

Types of cell

There are two types of cells used to make batteries, which are called primary cells and secondary cells. They are discussed as follows:

Primary cell

A primary cell can never be recharged because the internal chemical reaction cannot be restored, so a battery made up of primary cells is not rechargeable.

Secondary cell

A secondary cell, and it is also known as a storage cell, can be recharged because the chemical reaction is reversible. Dry cells which are made up of primary cell have a moist electrolyte that cannot be spilled whereas; the sealed rechargeable cells are secondary cells that contain a sealed electrolyte which can never be refilled.

Common Types of Primary Cells

The different types of primary cells in use today as described as follows: carbon-zinc dry cells that is one of the most popular primary cells, it is usually used for type AAA, AA, C, D. The negative electrode of dry cell is made of zinc. The positive electrode is made up of carbon. The output voltage given by such a single cell is about 1.5 V. Performance of the cell is better with intermittent operation.

Alkaline cells

Alkaline cells are another popular type, and it is also used for type AA, C, D. It has the same voltage as a carbon-zinc cell which is 1.5V, but they are longer-lasting. Its anode is made up of zinc, and the cathode is manganese dioxide in an alkaline electrolyte (potassium hydroxide). It can work with high efficiency even if it is continuously used because it has a very low internal resistance.

Zinc chloride cells

Zinc chloride cells are also referred to as a "heavy-duty" type battery. It is because of the modified zinc-carbon cell. It has a little amount of liquid leakage because the cell consumes water and also has chemically active materials. At the end of the useful life, the cell is always dry.

Mercury cells

Mercury cells consist of a mercury compound cathode, zinc anode, and a potassium or sodium hydroxide electrolyte. It is considered obsolete due to the hazards associated with the proper disposal of mercury.

Silver oxide cells

Silver oxide cells consist of a silver oxide cathode, a zinc anode, and potassium or sodium hydroxide electrolyte. It is also typically available as 1.5V, miniature button form. It is mostly used in hearing aids, cameras, and watches.

Common types of secondary cells

The different types of secondary cells in use today, as described as follows:

Lead-acid wet cell

Lead-Acid Wet Cell is a widely applied type of secondary cell which is used extensively in vehicles and other applications that requires high values of load current. The cathode is made of lead peroxide. The anode is made of spongy lead metal. The electrolyte is sulfuric acid. The voltage output given is about 2.1 volts per cell. Lead Acid Wet Cells are typically used in series combinations of 3 (6-V battery) or 6 (12-V battery). Lead-Acid Wet Cell that is used in vehicles has a reversible chemical process. For example, the battery produces lead sulfate and water when it reacts by producing current flow in an external load circuit. The battery produces lead, lead peroxide, and sulfuric acid by reacting to a reverse current from an external energy source.

Nickel-cadmium cells

More additional types of secondary cells are Nickel Cadmium (NiCd), cells that deliver high current, and they can be recharged many times. It can also be stored for later and effective use for long periods of time.

Applications of Nickel Cadmium (NiCad) cells and batteries

· Portable power tools.

· Portable radio and TV equipment.

· Alarm systems.

The electrolyte used in this cell and batteries is potassium hydroxide (KOH), but it usually does not appear above, as its only function is to act as a conductor. It is there for the transfer of the hydroxyl (OH) ions. The gravity of the cell and battery does not change with the state of charge.

Nickel-Metal-Hydride cells

Nickel-Metal-Hydride (NiMH) Cells are used in long-running battery performance. They offer higher capacity than NiCad cells, which is 40% more capacity. They only have a different negative electrode, while all the other components are the same as of NiCad cell. They are more expensive than NiCad cells because they can self-discharge more rapidly, and cannot be cycled as frequently as NiCad cells.

Nickel-Iron Cells

Nickel-Iron (Edison) cells were once used in railway applications and industrial trucks. They are now considered almost obsolete due to lead-acid batteries. Nickel-Zinc Cells were also used in some railway applications. Their application is in the electric car due to high energy density. They have limited life cycles for charging.

Fuel Cell

Fuel Cell is actually an electrochemical device that converts chemicals, such as hydrogen and oxygen into water and produces electricity in the process. As long as the reactant's hydrogen and oxygen are supplied to the fuel cell, it will produce electricity continuously and never go dead, unlike conventional batteries. The fuel cell uses methanol, and oxygen is being developed. They are mostly used extensively in the space program as sources of DC power. They are very efficient as they are capable of providing hundreds of kilowatts of power.

Solar Cells

These are the type of secondary cells that convert the sun's light energy into electric energy. They are made up of semiconductor materials. They are specifically arranged in modules that are assembled into a large solar array as they have to produce the required power.

Lead-acid batteries, lithium-ion, and saltwater batteries are a tested technology that has been used in off-grid energy systems. They have a relatively short life and least expensive options for the energy storage sector. Lithium-ion batteries are lighter, long-lasting, and more compact than lead-acid batteries. They also have a longer lifespan as compare to lead-acid batteries. But the lithium-ion batteries are more expensive than the lead-acid batteries. Saltwater batteries don't contain heavy metals; they only rely on saltwater electrolytes and there is no specific requirement for its disposal. While the batteries that use heavy metals, such as lead-acid and lithium-ion batteries need to be disposed of with special processes, a saltwater battery can be easily recycled. Still, at the same time, the saltwater batteries are untested. Through the series and parallel connected cells, a voltage higher than the voltage of one cell can also be obtained by connecting cells in series. The sum of the total individual values for each cell leads to the total voltage available across the battery of cells. Parallel cells, as compared to series cells, have the same voltage as one cell but have a more current capacity. The cells can be connected in series-parallel combinations to have a higher output voltage and more current capacity; the combination of series and parallel cells is called a battery.

The cells are connected in parallel combination for the higher current rating. Cells can be connected in a series-parallel combination to get a higher output voltage and more current capacity. For normal service life, it is very important to note the current rating of batteries, or even any voltage source, which is only a guide to typical values that are permissible. The actual amount of total current produced when the battery is connected to a load resistance is equal to Ohm's law.

This use of such batteries in solar cells means that the output voltage varies with charge and load much more than other batteries. As compared to the solar cell, if you are using an inverter, the inverter needs to be designed with some specific voltage swings. You may not be able to use Ni-Fe's which are used in the solar cell if your system depends on a stable voltage, such as, for example, the light intensity fluctuates if a person is running certain common DC appliances directly off the batteries and also when using Ni-Fes to power DC lighting. One can use a voltage regulator for electric appliances to feed those appliances, but that would decrease the efficiency even more.

The battery size also matters when buying a solar battery. It means that the energy storage capacity of the battery should be higher. The physical size of the battery also matters for the space one has for installation, but the capacity is most important. One more aspect to look for is power rating. A battery with low power rating and high capacity will only power a few appliances, but for a long time.

Similarly, if the battery has a low capacity but high rating power, it will power most of your appliances but for a shorter period. Thus, one should look for a battery with high capacity, as well as a high power rating. That can power multiple appliances for more extended periods.

Chapter 13

Solar power for homes

The sun is the most significant source of energy that fulfills our energy requirements with minimum or no disadvantages. It is estimated that the whole world is using only one percent of this considerable source of energy. If we start utilizing this source for most of our needs, we can make our society more developed. In this chapter, we will discuss the uses of solar power at homes. There are a lot of household systems that can be run through solar energy, which is comparatively inexpensive and durable.

The power we obtain from the sun to produce electrical energy is called solar power. The device which is used for this purpose is called solar panels or solar system. The primary purpose of this device is to convert solar energy into electrical energy. This solar energy helps in the production of photovoltaic power (PV) or solar power (sp), which are used for many purposes, e.g., used in the home for providing electricity, used for heaters, etc.

Ways to attain solar power at homes

Innovations in the fields are becoming more advanced. The scientist tried to invent different solar devices to fulfill our everyday needs. Here are some ways through which we can attain solar power as an alternate of electric power:

1. Photovoltaic system

2. Solar water heating system

3. Solar power plants

4. Passive solar heating

One of the best ways to attain solar power at home is to utilize solar energy through the solar panel. It is beneficial in a way that first, you have to buy it at a high cost, but once you have it, it saves your electric bills. The solar power system that is used to produce electricity in homes is called a solar panel. Sun is the natural source of energy. It provides units of light called photons (which are small packets of power). That travel from the sun to earth can be covered in just about 8.5 minutes, which is very short in covering such a long distance. These photons produce enough solar energy and fulfill the significant and necessary demands of our planet.

A report made by the international energy agency in 2017 highlighted that the growth of solar energy usage increases as compare to all other fuels. It is because of its beneficial importance. In the future, it is expected that people of our planet will depend on solar energy for all our energy needs.

Parts of the solar panel

The composition of the solar panel is complicated, but it makes our lives simple. The solar panel is made up of small parts. It is composed of tiny solar cells. These solar cells are made up of silicon. There are two layers of solar panels. One layer is positive, and another layer is negative, which combine to form an electric field just like in a battery.

Procedure

First of all, when you fixed the solar panel on your rooftop, there are a few steps to follow to be able to create electrical energy for your home use. At first, sun rays fall on the solar panel that produces direct current (DC) electricity, at which electrons flow in one direction within a circuit. It must be converted from direct current (DC) to alternating current (AC) to make it usable for a home. In alternative current (AC), electrons move in a random direction. For this purpose, the solar inverter is required. The solar inverter converts direct current into the alternate current electricity use by our home. From inverter, AC electricity moves to the breaker box. After that current passes to all types of electrical appliances at home, the electricity that is not used at a time shifts back to the electric grid and meter. This procedure is called net metering.

If the solar panel produces too much energy more than your usage, this extra amount of power moves back to the electric grid. It will help its users in utility bills. This whole process is named as net metering. If your home requires more energy than energy produced by solar panels, you can borrow it from energy credits to reduce the bills of electricity. The more solar panels people have, the more energy they produce.

How are solar panels installed?

Steps of installing solar panel at home

For the installation of the solar panels, there are some steps to follow. These steps must be performed in proper order and sequence. There are five main steps to install a solar panel at home.

1. Visit of engineers

2. Permits and documentation is essential

3. Order important equipment

4. Installation of the solar panel

5. Approval and interconnection require

1. Visit of engineers

You have to make a solar contract with an engineer and installer. The engineer should be capable of understanding the requirements of the installation when you complete your solar agreement. An engineer will then come to your house to check the structure and condition of your roof and to check the electrical status of the home. They will check your solar panel and check the grey box mostly affixed in the basement. They check it is there any need to upgrade or not. Sometimes they only check measurements taken by the installer.

2. Documentation

The second step comprised of permission and documentation through a proper procedure. The most important thing you need to apply for is the state and solar incentives. E.g., PACE, SRECs (solar renewable energy certificates), and PACE (clean energy financing certificates). After that, you need other building permits. These permits are related to the location of the area where you live. You can install the solar panel on the entire roof surface. But certain states allow installing the solar panel with 3 feet clear surface around. Most of the time, the paperwork is filled out by installers who know the policy of the state. You must take precautionary measures while doing paperwork. A contract with the appropriate person is a crucial step. The skills of installers matter a lot.

3. Order for important equipment

The third step is to order the required equipment for the installation process. For this purpose, you have to order two essential types of equipment, i.e., solar panel and solar inverter. You have to consider the quality of the solar panel and inverter. Installer recommends the brand of your equipment. After selecting the desired equipment, you will get your order of shipment on the day of installation. Be careful in choosing the right equipment.

You could be delaying your installation by making the wrong decision of selecting equipment and installer. In short, your equipment should be reliable and your installer should be punctual and skillful.

4. Installation of solar panel

It is the most important day for the solar homeowner. First of all, you have to be mentally and physically ready for installation. Your availability is necessary so that you keep your eye on the whole procedure. Installers start their work by making sure that the roof is stable or not for solar panel installation. After that, they will do electric wiring and connect it with a general power system and electrical panel. When electrical wiring is done, then racking is made by installers to support solar panels (the only part that's attached to the roof). Then they place solar panels on the racking. To convert direct current (DC) into the alternating current (AC), they will attach your inverter with the solar panel. The time of installation depends upon the three (3) factors; one is the availability of the installer. The busier installer is more time he takes. Second is the size of the solar panel system. Larger size takes more time than the smaller size. Another factor that increases your installation time is the process of net metering, which requires the installment of the power meter. So, the total installment of solar panels may take 3 to 4 days.

5. For approval and interconnection

It is the final step for the installation of a solar panel. It is last but not least. You have to be careful about the requirements of this step. Before connecting your solar panel with the electrical grid, you need the approval of a representative of the government who will check the whole system. He will check work done by the installer. He will check to assure that electrical wiring is done correctly. He also checks whether the installment of solar panel fulfills the standard codes of electrical and roof.

• This inspection then grid interconnection is made. The representative of your electric company does the final evaluation of the solar system. After this, your solar panel system is connected to the grid. It is the final step.

• Panels are used in houses for running machinery and electrical appliances. Solar energy is widely used in solar housing systems because it is inexpensive and cheap. With the small expenses, we can install a solar panel through which we can run motors, light bulbs, heating systems, etc.

• Uses of solar panels are increasing day by day.

• Persons living in rural areas, where electricity is not available, are getting more benefits from the solar panels.

• Installation of a solar panel is a complex procedure that includes five significant steps.

• It cannot neglect any single step because all are important and must be appropriately followed.

• With an appropriate engineer and installer and order, proper equipment is the critical step.

• The day is significant.

• The whole procedure of installation of the solar panel takes five to six days.

• You have to take precautionary measures for the installation.

Chapter 14

Solar power roof tile

Expectations of the energy performance of the buildings are continuously increasing. At the same time, everyone wants to avoid and move away from fossil fuels and nuclear energy and moving towards more renewable energy resources. An integrated solar roof that converts clean power and unlimited solar energy into electrical energy is now the main focus. In recent years, many amazing gadgets and devices are available, which utilize the sun's energy, and it's easy to overlook the most judicious use of the solar energy system that is rooftop solar. In broad-spectrum, solar energy can be used to fly an aircraft and charge a battery. Similarly, if we look at the narrow way, it can also be used to save homeowners (thousands of dollar each and every year) by cutting their energy use, carbon footprint, and utility bills.

Construction of a solar roof tile

A solar roof tile should be made up of high-quality construction elements that should offer sustainable weather protection, and it should also enable efficient energy production all over the roof's surface. Proper installation of solar energy systems helps to save you energy. Solar cell efficiency is the portion of power in the form of sunlight that can be converted via PVs into electricity. The effectiveness of the solar cells used in a solar energy system, when they are in combination with latitude and climate, determines the annual energy output of the whole system.

Benefits of solar roof tiles

Solar roof tiles can help reduce the total heat transferred into your attic by around 50% when we compare it to a direct nailed asphalt shingle roof. There is a specific trade key of every company that respects the intellectual property, trade secret, copyright, trademark, or any other personal or third party rights and expects the same from others. These tiles are made up of a flexible material, so they are easy to apply when doing the roof.

The tiles are also made from a lightweight recyclable plastic that is molded together with a flexible solar cell. The flexible solar panel design consumes less energy; it reacts to a broader spectrum of light. The solar roof tiles are made out of a lightweight recyclable material that is molded altogether with a flexible panel. The tiles can also be networked together to leverage the power of the sun. There is a specific built-in blocking diode technology; this charger is designed to protect against battery discharge at night. This device can also be used to maintain the charge on any 12-volt battery. If an input supply is unable to provide enough power to satisfy the requirements of a solar roof tile charger, the supply voltage will collapse.

• A minimum solar roof tile operating supply voltage can be programmed by monitoring the total supply through a resistor divider, so that the desired minimum voltage is achieved. The acceptance of the solar cells would allow them to charge even indoors, an essential consideration since portable electronics are not typically exposed.

To develop a solar roof tile, it is necessary to understand current roofing practices and materials. The tiles are laid on wooden battens that run along the roof and are attached by special clips or nails. The battens are, in turn, nailed onto the rafters of the roof structure. Sparking felt provides a secondary level of protection, which is laid between the rafters and the battens. This material offers insulation, prevents dust and air entering the roof space, and provides an extra waterproof barrier.

Most efficient solar system

• The most efficient solar system has 3.3 KWP. Such a solar PV system consists of 228 polycrystalline modules integrated into roof tiles. The 114 modules (sun-slates) are then connected in series to generate 400 V DC at their rated voltage under STC. Two strings, each having almost 114 sun-slates, are then connected in parallel to keep the voltage at 400V and also into one inverter of 3.82 kVA capacity. Then the AC output from the inverters is connected to the single phase of the three phases of the grid. So, the electricity produced by the PV system is fed directly into the grid. A communication cable is then installed to record electrical data (power, voltage, current, power factor, etc.), and later on, a weather station would be installed to record meteorological data (radiation, temperature, wind speed, rainfall, humidity, etc.). However, there are zero sensors installed to measure the temperature of the PV modules. The exposed width and length of the tile are known as the gauge and cover. The annual net electricity generated from this solar PV system was also monitored and recorded (in May 2009).

For the total calculation of the life-cycle energy use, emissions, and cost, an average annual electricity generation of 2 008 kWh/kW is calculated and used. The average net conversion efficiency of the solar PV roof tile system under STC is 11% based on the manufacturer's specifications, while the measured efficiency of the inverter is about 92%. The installation of a 3.3 kW PV roof tile system. The most important aspects of being considered in installing solar roof tiles are as follows:

- PV array

It converts sunshine into electricity (both direct and diffuse radiations), so it works even on the cloudy days.

- Inverter

It converts the DC generated by the rooftop system into alternating current that can be used by the household or fed into the grid. The inverter converts (DC) into alternate current (AC) so that the energy produced by a solar PV-module can be used to power your home's electricity. This process is not 100% efficient. While inverter efficiency technology is improving, it can still account for as much as an 8% inefficiency.

- Meter

Especially for grid-connected systems, spare electricity generated during the day flows out to the grid and is sold to the local electricity supplier. Fitting an electricity meter will measure electricity fed into the grid.

- Battery

For off-grid systems, batteries are used to store electricity generated by the rooftop system during the day to use when needed. If one has a hybrid system with a solar battery, the battery will not operate at its rated wattage. The inverter efficiency for batteries is in the realm of 92%, representing only an 8% loss.

- The solar roof tiles are fitted directly into the sub-roof structure using a fitted sub-construction made of stainless steel or wood. The weight and size of the solar tiles are designed in such a way that it takes only one roofer to lay them. The connectors of the electrical wirings on the back of the solar tile ensure foolproof and safe cabling, both of which can be done at the same time as the installation of the roof. Small and lightweight solar roof tiles are easier to install than regular solar panels. The installation of solar roof tiles is also easier and does not require any additional types of equipment because they don't require a roof evaluation or any extra space. A sloping rooftop is an ideal site, and that's because modules are mounted using frames, but a flat rooftop can also be used for maximum power output.

Key Factors

Roof conditions vary greatly, and several key factors should also be considered when mounting a PV system onto a roof. Some of the most important and significant of these factors are the following:

- The direction that solar roof tile face, and the angle that they are mounted, also affect the output of the panel/tile/slate. If they are not positioned at an optimum angle and a direction, then a conventional PV solution will not then produce rated output for that location. It should be considered a must when accurately calculating the output of solar energy.

- Direction - The direction a solar PV-module should ideally face for optimum sun exposure varies depending on location. Hence, the direction solar installation face, along with their angle, or tilt, at that orientation. They are said to be vital inputs to calculate the solar power potential for a property accurately.

- Geographic orientation - northern or southern hemisphere: PV systems maximize power output on south-facing roofs in the northern hemisphere and north-facing roofs in the southern hemisphere (up to

95% efficiency). For best results, the roof pitch of the property should ideally always be equal to the latitude angle, with the PV solution mounted at the same angle as the roof.

- **Azimuth:** The path length of the sun through the atmosphere increases with the increase in the azimuth angle.

- **Tilt/angle of inclination:** The angle from the horizontal formed by an inclined roof or mounted PV system on a flat surface.

- **Available area:** when there is a larger surface area available, the greater the power potential. Systems can be small to fairly large. For a grid-connected system, the required area could range from 8meters to several hundred square meters.

- **Shadowing:** Tall trees or neighboring buildings shouldn't shadow the roof. Even if there is slight shading, it can cause a significant loss of energy.

- Rooftop systems can be roof mounted. PV modules are fixed on frames above the existing tiles or integrated with the roof. PV modules are also an integral part of the building structure, replacing conventional roof tiles in new buildings or re-roofing.

Continual improvements in cell efficiencies, coupled with reductions in production cost through new processes and economies of scale, will enable PV to become an economic reality for urban power generation soon. Solar roof tiles have the potential to bring PV to the largest market. Perfect design enables PV to be integrated in an aesthetically pleasing manner. The solar roof tile represents the combination of the mature tile industry, with good quality plastic and photovoltaic technologies. The solar tiles are screwed directly in the sub-roof structure using a fitted sub-construction made of wood or stainless steel.

Also, the size and weight of the solar tiles are designed in such a way that it all takes only one roofer to lay them. The electrical connectors on the backside of the solar tiles ensure foolproof and safe cabling, which can be done, both at the same time as the installation of the roof.

Chapter 15

Solar-powered air conditioners

The climate is changing. It is becoming hotter day by day. The temperature of Earth has increased due to global warming and the greenhouse effect. Man is always in search of creating luxuries and comforts for him. Man found a way of fighting this unbearable heat and ambient temperature. Air conditioners are used to reduce the effect. Air conditioners are used everywhere.

An air conditioner is a machine that works by the cycle of refrigeration. Warm air is discharged and replaced with colder air. It fixedly treats the air, mostly encapsulated in an area. This mechanism makes the room's temperature cooler than the outer environment. The air conditioner is always attached to walls. It is not placed in the center of the room. If it is done so, the warm air will not leave the room, and the room gets warmer instead of getting colder.

Air conditioners make your life very luxurious, but the ordinary air conditioners consume a lot of electricity. So, solar-powered air conditioners are introduced. They work using solar power.

Types

There are two types of solar-powered air conditioner. These are:

Hybrids

Hybrid means a combination of two. These systems combine photovoltaic current (PVC) and direct current (DC). It means that this system can work both on photovoltaic current (PVC) and direct current (DC). This system switches automatically from DC to PVC when required. During day time, when the system is in hybrid mode, it will charge the batteries. The system will consume the energy from the battery when not in hybrid mode, for example, at night.

Solar hybrid air conditioners take the improvements made by inverter technology to the next level. It captures thermal (ambient) energy. The work done by the condenser is reduced significantly, and this system also extends the time the compressor completes its cycles.

Solar-Powered Evaporative Coolers

It is also known as absorption chillers. This system uses the processes of condensation and evaporation to cool and heat the water present inside them. Fans and motors are used for cooling and heating, respectively. These fans and motors are powered by solar energy. These chillers can function for several hours in the absence of electricity.

As absorption chiller works by water, it will reduce the power consumption, but increases the water consumption. If we contrast and compare both the above solar-powered air conditioners, the consumption of water makes hybrid systems better a choice than chillers as it does not increase the consumption of any particular entity.

The Basics on How Solar Air Conditioning Works:

Principle

The solar air conditioner works on the law:

"A gas expands on heating and contracts on cooling."

For example:

There are two cylinders, cylinders A and B, with helium inside them. Cylinder A is heated, and cylinder B is cooled. The volume of gas in cylinder A will increase while the volume of gas in cylinder B will decrease.

An air conditioner consists of a refrigerant (gas), compressor, heating source, evaporator, and condenser. The air conditioning systems use a cycle for the refrigerant (gas). The compressor (within the condenser) works to pressurize the refrigerant and enable the method to initiate. An external heat source is used to apprehend ambient temperature and grant the heat to the refrigerant, and this process maintains pressure within a connected circuit of refrigerant. It is the general principle of the air conditioners.

The solar air conditioners use solar plates. These solar plates capture heat energy from the sun and convert it into DC. These plates are placed adjacent to the compressor. The heat is provided to refrigerant (gas) before the compressor. Factors that make the plate more efficient:

1. The heat-conducting coils

The flat collector plate has heat-conducting coils inside it. These conducting coils increase the conduction of heat.

2. Heading tank

The heading tank is full of water; it is also present inside the plates.

3. The surface area of plate

The larger the surface area of the plate, the more heat it can capture.

These additions optimize the efficiency of the collecting plate.

Compressor

The compressor is operated through electricity. It is called the core of the air conditioning system. It pumps refrigerants into the system. The main purpose of a compressor is to compress refrigerant vapor to high pressure. The compression makes the gas hot for the evaporation process.

Refrigerant

Refrigerant is a substance that performs as a cooling agent. It absorbs heat into the system and circulates inside the circuit of the air conditioning system. An expansion device is located between the condenser and the evaporator. It controls the amount of refrigerant flowing into the air conditioning system.

Evaporator

The evaporator uses the refrigerant in a liquid state to absorb the heat from the space. It is cool in the system for the circulation process of the gas. The evaporator is located inside the indoor unit present in the cooling area.

Condenser

The temperature of the refrigerant is low at the condenser because the heat from the system is removed. The condenser is also present inside the outdoor unit along with the compressor.

Photovoltaic, which is a combination of two words - photo for light and voltaic for electricity - converts the energy of sunlight directly into electricity. The conversion from the sunlight into electricity is occurred because of the PV effect. The term "solar cell" is reserved for many devices intended specifically to capture energy from sunlight, whereas the term "photovoltaic cell" is also used when the light source is unspecified.

The energy from the sun is captured by the PV cells and changed into electrical energy. The charge controller then regulates the electrical energy. The electrical energy coming from the PV cells is in DC. It is needed to be converted into an alternating current by inverter will. It is done as the compressor needs AC to function. The working of the system commences when the cold, low pressure (refrigerant) runs beyond the evaporator loop inside the cooling space to incorporate heat. The cold liquid that ran into the evaporator loop comes out as a low-pressure gas. This low-pressure gas is conducted outside and contracted by the compressor to shift to hot, high-pressure gas.

The hot gas is moved into the condenser coil and releases its heat as the outdoor air blows over the coil. It causes the hot gas to condense back into a hot liquid. The hot liquid is transported back to the evaporator by moving through the augmentation device, which lessens the ambient temperature and pressure of the liquid.

This process promotes energy maintenance of up to 30 - 50%.

Once the room reaches the aspired temperature, the condenser will veer off. The refrigerant continues to move in a cycle until the room temperature differs too much from the aspired temperature.

Throughout this method, constantly, the pressure within the circulating refrigerant decreases and hence reduces the capacity for the system to keep the aspired temperature. When the pressure sinks very low, the compressor starts working again. It competently re-pressurizes the refrigerant. Counter to this, the heat imparted from the plate collector will pressurize the refrigerant; therefore, the compressor remains off for a longer period. The compressor and solar heat collector are paired in such a way that the system can operate with a smaller compressor relative to a comparably sized inverter system.

The electric air conditions can have adverse effects on health and the environment. It creates noise pollution as it causes a lot of noise. It can affect our eyes as causing conjunctivitis, blepharitis, and other diseases of eyes. It causes dry skin. It also causes respiratory tract infections. It produces chlorofluorocarbons (CFCs), which ultimately damage the ozone layer. The consumption of electricity is, of course, is demerit of electrical air conditioners.

Solar-powered air conditioners can be used as an alternative to reduce the effects as mentioned above of air conditioners powered by electrical energy.

Advantages of solar-powered air conditioners

Sun is a reservoir of the gigantic amount of energy. The energy from the sun is renewable, so it should be utilized. Different objects made solar-powered can save the world from the energy crisis.

Solar-powered air conditioners reduce electricity costs. It has a very low grid demand.

Solar powered air conditioners can be used without electricity. Having fewer power outages is one of the benefits of solar-powered air conditioners.

These air conditioners are environmentally friendly. It saves the environment from hazardous gases and pollutants, which equally affect our environment and our health. Solar-powered air conditioners use has reduced greenhouse gas emissions.

Dis-Advantages of solar-powered air conditioners

The solar panel can be very costly at first, but once bought, it doesn't need any additional cost. It charges using solar energy.

The period of working without electricity is less, but it can be increased.

Chapter 16

Solar-powered vehicles and boats

Have you ever heard about solar cars and boats?

Like solar panels on the rooftop, All the solar vehicles are run by electricity produced by solar panels that are placed on the upper portion of that car. These solar panels consist of photovoltaic cells that change solar heat energy into electrical energy. This electricity forces the battery to start the car.

Photovoltaic cells are composed of semiconductors like silicon, which collect sun energy when sunlight falls over the panel. The photons from sun energy strike the electrons and release them from their atoms. These free electrons create electricity that gives power to the battery and motor of solar cars.

Positive aspects of using solar cars include:

- A solar car reduces the utilization of fossil fuels.
- Solar panels are noiseless so no chance of noise pollution.
- It is environment-friendly because it doesn't cause any toxic air pollution

- Unlike gasoline engines that produce greenhouse gases. Solar panels do not create such hazardous gases.
- Electricity from thermal energy is all free of cost.
- Another benefit of the solar vehicle is it keeps the driver stress free from diesel oil. The solar vehicle has that ability that it can run without diesel oil.
- A solar car also reduces the risk of air pollution.

Let's discuss some negative aspects of solar vehicles that are listed below:

Solar energy is not available at night, neither on cloudy days. That's why heavy batteries are required for the storage of solar power.

Another drawback is solar-powered cars are very costly and less efficient.

Scientists are trying hard to minus all the negative aspects and make them more efficient. But sadly, solar cars are still unable to be utilized as daily transport. Moreover, mostly solar-powered cars created for racing purposes. Mostly all the solar racing cars are designed with storing consumed sun energy for the aim of racing.

World Solar Challenge is a popular racing tournament in Australia. The pioneer of the World Solar Challenge race is Hans Tholstrup. He created a solar car named the "Quiet Achiever." World Solar Challenge is a platform to promote solar car technology. It highlights the brilliant efforts made by the world wide best engineers and companies.

New advancement in the world of solar technology is building more efficient solar cars focusing on high power batteries that save space.

Some of the popular companies are also using solar panels on the cars and putting their efforts to create a hybrid solar car. Hyundai placed the solar system on an electric Sonata, which will give the power to run 22miles daily. Toyota is also putting its efforts to create a solar car that runs up to 35 highly efficient miles. Toyota placed solar panels over the roof of the car to run air conditions in cars.

The fastest solar racing car that was created by students belongs to the University of New South Wales. They named the car sun swift IV. But a solar car 100 percent working on solar energy is not developed yet.

General Motors invented the first-ever solar car. They showcased it in Chicago during 1955. This solar car consists of 12 selenium PV cells and a system attached to a small size motor (Pooley). The car was small in size. Electric pool motor forces the pulley to run the rear wheel.

Another solar car named "Sion" is well known for the first hybrid car that charged by itself depending on sunlight energy. Only 30 minutes are required to charge Sion for up to 80%. It can run 155 miles with 80 % charging on a daily bases. A German company, Sono Motors, invented Sion.

A project by the Netherlands is known as Light Year one, which was introduced in June 2019; Tesla and Ferrari designed it. The surface of the car was covered with solar panels. Light year one is the first solar car that runs 450 miles so far.

Hanergy is a Chinese solar panel development company that aims to create a solar car composed of lithium-ion batteries. According to Hanergy, 6 hours of solar energy must provide 8 to 10 KWH electricity that enables a car to run up to 80 km per day.

Why are solar cars not for daily use?

First of all, the solar panels in use nowadays are not very efficient. The efficiency level is about 15 to 20 %. The whole car required a lot of space and heavy batteries to consume and store a lot of sunlight energy, which consumes a lot of space and cost a lot that everyone cannot afford. The solar panels over the roof are expensive, and its heavyweight increases the weight of the car. So, that's why covering the whole car with solar panels and a heavy battery for high energy storage becomes highly expensive.

Solar world technologies are now working on developing lightweight batteries with highly efficient photovoltaic cells.

Cars for public use

In 2013 first solar car was created for family use. The Case Western Reserve University invented the solar car with the ability of faster charging.

Solar railways

Imperial College London declared that to give solar power to train the renewable traction power project is using trackside solar panels which provide solar electricity to the trains. Alongside Indian Railway System states that they will be using PV board to provide power to air conditioning systems in Indian trains. By introducing solar trains, they can save about 90,800 liters of fuel per train annually, which leads to a decrease in the levels of 239 tons of carbon dioxide.

In South Wales Australia, a solar train was introduced during 2017. The interesting thing about this solar train was that it consists of solar board panels and onboard electric batteries with the capability to take 100 passengers and can cover a 3 km path.

Hybrid airship

An Australian company aims to invent an air crane named Sky Lifter, which can fly upward vertically and have an ability to carry 150 tons of loads and fully powered by the solar panels. It is a project designed as a delivery aircraft

A Canadian company is working on a solar hybrid airship that can completely rely on solar power and have the ability to travel anywhere in the world without using any fuel. To transport cold medical supplies and other equipment's to Africa and Canada.

Solar spacecraft

Seattleites and other space crafts are also empowered by solar panels, which can provide long-lasting power without utilizing fuel.

Communication satellites are composed of multiple radio transmitters that constantly operate throughout its lifespan. It's quite hard to operate a satellite with batteries that rotate in orbit the whole year.

Solar Water ships

Marine solar panels are fixed on the surface of the boat, and just like the solar panels in homes and cars, sunlight energy is converted into electrical energy, which gives power to the boat. The remaining energy can be later used for night.

Solar installation on a boat in a proper way is really important.

While installing panels, make sure that the position of the solar panel is in the right direction toward the sun; otherwise, the boat will not work efficiently if the inclined direction is not right.

Try to select appropriate solar panels for your boat. Flexible solar panels will be appropriate for your boat but are highly expensive as compared to rigid panels that are more affordable and easily manageable – the brighter the sun, the stronger the electricity to the boat. As long as the sun is shining over your boat, No need to worry about phone battery when the solar panels are fully charged; make sure to sail your boat according to the sunlight direction.

Marine solar system

A marine solar system consists of four (4) elements:

1. Solar panels
2. Charge controller
3. Inverter
4. Battery

When sunlight fall on the solar panels, it converts the sun energy into electrical energy that empowers the boat to run. The boat speed is lower during cloudy days.

The charge controller protects the batteries and cells from overcharging as overcharging damage the battery

The inverter is there to change direct current into alternating current. Depending upon the boat setup, you will need an inverter because household appliances (television, microwave) in the boat utilize alternating current. In contrast, batteries are required to store a high amount of energy for further use.

How many solar panels does a boat need?

The requirement of the solar panels is different for both the yacht and a sailboat.

The number of solar panels depends upon the size and type of the boat, the amount of electricity it uses, the amount of solar energy a boat receives. Some boats can ménage with 100-watt of solar panels, whereas other boats can ménage with multi solar setup.

Japan's Nippon Yusen KK, which is Japan's biggest shipping line, use solar panels that provide 40 KWH power, which should be affixed on the surface of the 60,213 tons of ships that are to be used by Toyota Motor Corporation.

Turanor Planet Solar introduced a yacht in 2010 named Catamaran, which is 30 meters in length and has 15 meters width. Solar panels of 470 square meters provide electricity to this solar yacht. It is the largest solar system boat created so far with the inbuilt function of circumnavigating the world.

Chapter 17

Solar Power for Laptop computers, tablet, and smartphones

Solar power has so many uses in our daily life. It makes our life easy and tension free. Many electrical appliances are now independent of the conventional electricity method because of solar power, which converts solar energy into electrical energy. Here are some of the applications of solar power:

If people don't have electricity in their area, or they are unable to pay expensive electricity bills each month. They should use the alternative of electric power, i.e., solar power. Now software engineers, university students, teachers, and freelancers can power and recharge their laptop batteries by use of solar panels only. The main purpose of this valuable book is to tell the readers about solar power and solar energy. Let's discuss some of the uses of solar power in laptops, mobile phones, tablets, and computers.

Structure and mechanism of solar power

Here is a brief structure of solar power kits use for charging electrical appliances. A solar kit consists of three (3) small solar panels that work together to give you a maximum amount of energy. A red light indicator is present at the front side, which senses sunlight either reaches or not to the panel. There are different straps present on the backside of the solar kit. Users can attach it with anything they want, e.g., bicycle, Scotty and motorbike, etc. A small hole is present at the backside of the solar power kit so that cable can easily come out from this hole, and the user can charge their mobile phones, laptops, and electric appliances outside of the solar kit. At the inner front portion of the solar power, the kit is the circuit box present between solar panels and huge inner battery. On the circuit box, there are two different cables attached. One cable charges the battery and attaches to it. The second cable is used to charge all types of electronic appliances. The power gained from the sun passes from solar panels to the circuit and then enters into the internal battery. With the help of this battery, people can charge their laptops and whatever they want. A battery contains input as well as outputs also. Input comes from the solar panel into the battery and output charges electrical appliances. It contains a USB port to charge the small devices, e.g., mobiles, cameras and cordless phone, etc. It consists of so many different adapters to charge all types of electrical devices, e.g., adaptors for laptops, cables for USB, and adaptor for charging the car as well.

The small five-light indicators are present. When all lights are on, it means that the battery is fully charged, and when lights blink, it means it searches for sun rays.

How this solar power work

It is very easy to check the working of a solar power kit. We can check the charging percentage of the battery by simply pressing the button present on the battery. The light indicators show the exact amount of charging. The more lights light up, then it means the more charge the battery has. If by pressing the button, only three lights light up, it means the battery is 65% charged. If two lights light up, it means the battery is 45% charged. Also, if no light is shown, it means the battery is dying, and it needs to charge.

Solar power for mobile phones

It is very easy for users to charge mobile phones with the help of solar power. Take a connecting cable to connect its one end with the mobile phone and another end with the USB output of solar power battery. It takes only a few hours or a few minutes to charge your phone fully.

Another method of charging a mobile phone is that put your mobile phone and battery together in the solar kit case. If you zip up the entire solar power kit, it will secure your Smartphone inside and also protect your mobile phone during an accident.

How to charge laptop computers with solar power

The following are the steps to connect your laptop computers with solar power. At first, Take a solar power battery and connecting cable that comes along with it and then take MagSafe adaptor to connect it with an output of solar power battery and connect its other end with your laptop. By doing this, your laptop starts charging. No, after this, your laptop is running on the solar power battery.

Solar power for tablets

You can operate your tablets on the solar panel, which gives solar power. A foundation name OLPC (ONE LAPTOP PER CHILD) give their services to the area where electricity supply is not available. This foundation provides solar power with new low-cost tablets. A tablet providing by this organization consists of solar power cover. This cover performs three major functions.

- It helps users by protecting the tablet.
- It supplies power to the tablet by solar cells integrated within it and a thin battery, which also helps in providing power.
- It fulfills all computing requirements of the tablet.

The solar power kits are also used to supply a large amount of power to the tablets and also help in charging tablets, which is used for job purposes, connecting with your close ones, and for completing assignments for school.

Some of the best solar power that is widely used for laptop, computer, and tablets are:

- Feel solar power bank
- Add to portable solar power bank
- Beartwo portable solar charger
- Rav power solar charger

Benefits of solar power

Solar power gives so many benefits to its users. It is a low maintenance cost, and it reduces electricity bills of the laptop, computer, tablet, and mobile phone users. The most important is that it is a renewable energy source. It gives benefits to its users as long as the sun is available. It reduces dependency on electricity.

Solar power for computer

Running a computer on solar power is very good. The computer operates very fast on solar power. It reduces the electricity bills of its users. The use of solar power is very affordable. The amount of solar power depends on the capacity of the computer and the need for the computer. Laptop computers use less amount of energy than desktop computers. But its amount depends on the hardware design of the computer. Following are some steps to connect your desktop computer with solar power:

Steps for connecting solar power with a computer:

1. At first, buy a solar panel whose energy must be according to the need of your computer. The small computer uses 40 to 50 watts of power, and a large computer uses 90 watts power, probably.

2. Buy a battery of 12-volts. You can use the battery of cars, deep-sea submarines, or even small batteries.

3. Buy an AC-DC inverter. It consists of a wall socket into which you can plug your computer.

4. Connect your solar panel to the charge controller.

5. Connect your charge controller with the battery. Connect the positive wire of the charge controller to the positive battery and negative wire of the charge controller to the negative battery side.

6. Before using for the first time, allow the battery to charge fully for several hours or even for a full day from the solar panel.

7. Seven steps are to connect the inventor with your battery with the help of battery clips. Connect positive battery clip to the positive side of the battery and negative battery clip to the negative side of the battery. Be careful while connecting.

8. Now switch on the invertors and connect your computer with it. After that, switch on your computer.

So, this is the whole procedure to charge your computer with the help of solar power panels. Solar power is used to reduce your electricity bills: people start to depend on solar panels instead of utility companies to run their electrical appliances, e.g., laptops, computers, tablets, and smartphones. It will reduce your bill by about 50%. You have solar panels in your town, home, and office, as well. It uses solar energy instead of electrical energy to power your electrical appliances except in the case when the sun is not present, such as cloudy weather.

Power helps in protection of electric appliances from high voltage

- The voltage of the electricity supply to your house changes due to the problem. It may cause a short circuit. Which, as a result, burn many electrical appliances and make them useless for the users. So. protect your laptops, computers, smartphones, and tablet from damaging use of solar panels in which voltage is fixed and cannot change like electricity voltage supply to your houses.

Power charges your phones, tablets very fast:

- Power charges your electrical appliances so fast. It maximum takes 2-4 hours to charge your laptops, desktop computers, and tablets fully. Charging also depends on the need of electrical appliances. Usually, smaller devices like smartphones, charge more quickly than larger devices like laptops or computers, which take a long time because of hardware components.

Power makes laptop users independent of load shedding

- Some areas of this world, such as those where electricity is not available. People face many problems that live in these areas. But now the modern era enables people to use solar power to charge their laptops, computers, and mobile phones. Now by the advent of solar power panels, people can easily perform their jobs on time. They can easily connect with their closeness independent of electricity availability.

Power runs so many computers at a time

- You have a solar panel system in your house or offices. This system utilizes a very large amount of solar energy and converts it into electrical energy. This large amount of solar energy runs so much computer system at times, which is connected with it.

Power charges your mobile even you are outside of the house

- The power which we get from solar power kits helps its users to charge their mobile even when they are outside of their house. The solar power kit provides energy only as long as the sun is available.

Chapter 18

Examples of Solar Power Projects

With the advancement in technologies, scientists are in the process of making things that use solar light. In this chapter, we will discuss some solar power projects. Some projects have been completed while some are in process.

BENBAN Solar Park, Egypt

BENBAN solar park project has been made to be the world's largest solar installation. As the name indicates, this park has several power plants installed by different countries as well as by industries. It aimed to produce almost 1.8GMW power of energy. For this project government has allotted 37.2km^2 of land in Daraw Markaz's town. About 41 plants are designed to plot on it. Photovoltaic panels are mounted on it that are fixed and have immovable frames. They are planted in a parallel pattern and are connected to inverters. These inverters used to convert direct current (DC) to alternating current (AC). This power is then transferred to the next nearby grid.

For manufacturing and other purposes, the water pipeline is used from the Nile River at a distance of 16km. Hence, the government has provided all the basic facilities for the successful making of a plant.

Several other solar projects are also in process in BENBAN solar park. The biggest solar plant in it is Infinity 50. This project is under the supervision of the Infinity Solar consortium. This consortium, with the collaboration of Phoenix Solar, is making three projects, i.e., Phoenix 50, BSEP 50, and MMID 30. A total of $190m is invested in these projects to get 166.5MW of energy as an outcome.

All the power that is generated by this giant solar park will be transmitted to several substations. EETC has constructed a high voltage carrying the capacity line (220kV) for power transmission to substations and then to the national electricity network.

LONGYANGXIA Dam Solar Park, China

LONGYANGXIA dam is subjected to generate 850MW of energy. It has been evident that there are about 4 million solar plants are planted. It is considered to be the world's largest solar farm built in China. State Power Investment Cooperation is responsible for its operation. £721.3m (6 be yuan) is invested in its manufacturing since 2013.

It has been evident that the enormous amount of energy generated by this giant dam is enough to run 200,000 houses. The major benefit of this is that the power generated by this plant is clean, and it gives power to the electric grid as well.

SUNGROW Solar Farm, Huainan, China

China has unique importance in the world due to its value. They utilize natural sources in such a way that blows one's mind. SUNGROW solar farm is one of the examples. It is the world's largest floating solar farm. It is located in the Anhui province of Huainan, a city of China. This floating farm has a variety of landscapes and is generated 40 MW of energy.

This farm is made on the surface of the water, covering a coal mine that was collapsed. One hundred sixty-six thousand panels are installed there. The energy produced by this farm is giving power to almost 15,000 homes.

The building of farms on floating and destroyed coal mine has many advantages. China is considered to be the worst populated country. This solar farm not only protected lakes, animals, and the environment but also cool the panels and protect them from overheating and ultimately maintain their efficiency.

China is now planning to generate an additional 150MW of energy by 2019.

Solar Star, California

Solar Star project is the largest solar power generation project in the United States. It has two dimensions: Solar Star 1 and Solar Star 2, located in two cities of California - Kern, and Los Angeles. It covers about 13sq.km of the land. Sun Power Corporation is investing money for its construction and operation. The project is subjected to generate 579MW of energy, which is enough to power 255,000 houses. About 1.7 million modules are installed on single-axis trackers. These are made of silicon and are monocrystalline.

Mike Fehr, general manager of the Solar Star project, shared the aim of the project. He says that they are aiming to generate 33 percent of the total power generated from renewable sources by 2020.

This project has not only economic benefits but also has environmental benefits – about $500 million increase in the economic budget. More than 650 jobs were given for construction purposes. The solar panels that are installed in it are of high efficiency, and the Oasis power plant technology reduces much of the land usage. BHE solar evidence the removal of 570,000 tons of CO_2 per year.

Solar Panel Bike Path, Netherlands

After the success of solar power plants everywhere, the Netherlands decided to accept the challenge. They made solar panels on the bicycle track. First, it was covering 70m, and then it is extended to 20m more. The photovoltaic panels that are used in it are covered with thin films. This path is producing 3000KW-hr energy that can run a house for one year. This idea is beneficial as it is a busy track, but it has not generated energy as was expected.

Solar Tindo Bus, Australia

Solar Tindo bus is a solar-powered bus in Australia. "Tindo" is an Aboriginal word used for Sun. It is the first vehicle that is powered by 100% solar energy. There are no solar panels in it. Instead, the bus received electric signals from the station. The bus provides services free of charge and is also offering AC and Wi-Fi services free of cost to all the passengers.

Unlike other buses, this unique bus has no combustion engine at all. It is pollution-free and saves 14000 liters of diesel as well.

Canal Solar Power Project, India

Due to the rise in temperature, evaporation becomes the major problem of India. To protect canals, covering a thousand miles, from evaporation, the government of India planned to build solar panels over the canal. This idea not only saves them from the loss of water but also take their economy to the next level. It is covering 750km of canals and producing 1MW of energy. The panel is limited, but if they extend it, then it will cause a big change for them. The completion of this project can prevent almost 90 million liters of water from loss due to evaporation.

Photovoltaic Road, China

China has constructed a road made of solar panels. This road costs $11 per square, although it costs much, not more expensive than solar panels. The road is so strong that it easily bears the weight of small trucks. The Chinese are very hopeful about this project.

The Crescent Dunes Solar Project, Nevada

The crescent dune solar power project is constructed near Tonopah, Nevada. The project started in August 2011. At first, there were limited resources, and ultimately the construction process was slowed. The project was then completed around 2014. This project has a central receiver tower that makes it different from others. The tower receives solar rays from the Sun and then uses it. Another importance is its storage. This solar plant is enriched with the world's largest storage capacity. The plant is aimed to produce half a million kilowatts of energy. It also reduces the emission of 290,000 million tons of energy per year.

Tokelau Renewable Energy, Tokelau

Tokelau is a very small piece of land, but it is the only land that is dependent on solar energy. This solar project generates energy that is most likely enough to power all houses. Due to this project, the neighboring island of it is also working on the same project. It helps them to reduce the air population by avoiding discharge of 2.5 million pounds of CO_2.

Solar impulse

Nothing is impossible when you have a passion for doing it. Solar impulse is the best example of traveling in the air by solar power. The plane made for this purpose has wings larger than the jumbo jet but only have weight about 5000 pounds. The first flight taken with the help of solar power was in December 2009. The next flight, i.e., Solar Impulse 2, covered the entire world only with solar power. The main aim of this is to take maximum advantage of natural resources, especially the sun, that has the most powerful and enormous energy. This project was not only using solar energy but also finding ways to remove pollution and other harmful processes.

ARCO SOLAR, California

Arco solar is another solar project in California. This plant is subjected to produce 5MW of energy. Silicon wafers are used in these plants that directly receive solar energy and use it to produce electricity. The pattern in which the solar panels are installed is assembled in a way that the trackers follow the sun in both azimuth and elevation.

VANGUARD 1

Vanguard is the first satellite that orbits around the earth that operates on solar energy. It is designed to test the three-stage vehicle and the environment's effect on satellite and its system. Although it is the second satellite, yet it is the first that uses solar power cells. It is still working.

Chapter 19

Conclusion

Solar power is the only renewable energy that is limitless and cannot ends. Scientists are trying their best to manufacture the technologies that not only use solar energy but also store this miraculous energy for future use. Many countries get victory in this regard, and other countries are in progress.

Many new inventions are being made in the field of renewable resources, and also many different gadgets and technologies are taking leverage of solar power in the modern era. The list and line of examples of solar energy are endless, whether it's powering renewable transportation or charging a wireless speaker, and it is providing more options for homeowners who decide to "go solar." Solar energy is mostly used to power automotive, lights, pools, heaters, and gadgets.

Although the solar-powered products are increasingly complex at the same time they are providing to with much ease for examples: solar transportation, solar tech, solar lighting, solar heating and rooftop solar and many other such as;

- Tablets

- Cell phones
- Music speakers
- Wearables
- Thermostats
- Solar air conditioning
- Solar visor radios
- Rechargeable flashlights
- Solar dryers
- Freezers/mini-fridges

The world is also moving far away from fossil fuels and seeking renewable sources for everyday electronics, so consider solar as the best option. It will continue to be adopted by the masses for any product, and that can be exposed to sunlight. The ultimate goal: as the conversion efficiency of solar polar is being improved, this trend will also be expanded into new markets and new products. Through solar power, there is decarbonization of the energy sector and the reduction of carbon emissions, which will limit climate change. As an improvement in the solar energy sector and also the reductions in the costs of solar energy systems as it is a fast-evolving industry in recent times, the social energy industry is also changing rapidly through its innovations along the entire value chain and further rapid cost reductions. Different pathways are being introduced for the global energy transformation. There is also scaling up of solar PV capacity so that the transformation can be fast and pure. In many different countries, several research projects and prototypes are ongoing, which are definitely increasing future market growth by exploring innovative solar technologies. It is highly beneficial for the industry as it increases socio-economic benefits. More advancement is being announced for solar power systems, which can regulate themselves when there is a slight change in climate and can generate more electricity even when there is a very low intensity of sunlight. The capacities of solar energy systems are being improved so that it enables the transition to a low-carbon, sustainable energy future.

Printed in Great Britain
by Amazon

67072496R00088